Human–Computer Interaction Series

HCI is a multidisciplinary field focused on human aspects of the development of computer technology. As computer-based technology becomes increasingly pervasive – not just in developed countries, but worldwide – the need to take a human-centered approach in the design and development of this technology becomes ever more important. For roughly 30 years now, researchers and practitioners in computational and behavioral sciences have worked to identify theory and practice that influences the direction of these technologies, and this diverse work makes up the field of human-computer interaction. Broadly speaking it includes the study of what technology might be able to do for people and how people might interact with the technology. The HCI series publishes books that advance the science and technology of developing systems which are both effective and satisfying for people in a wide variety of contexts. Titles focus on theoretical perspectives (such as formal approaches drawn from a variety of behavioral sciences), practical approaches (such as the techniques for effectively integrating user needs in system development), and social issues (such as the determinants of utility, usability and acceptability).

Titles published within the Human–Computer Interaction Series are included in Thomson Reuters' Book Citation Index, The DBLP Computer Science Bibliography and The HCI Bibliography.

More information about this series at http://www.springer.com/series/6033

Fang Chen • Jianlong Zhou • Yang Wang
Kun Yu • Syed Z. Arshad • Ahmad Khawaji
Dan Conway

Robust Multimodal Cognitive Load Measurement

Springer

Fang Chen
National ICT Australia (NICTA)
Sydney, NSW
Australia

Yang Wang
National ICT Australia (NICTA)
Sydney, NSW
Australia

Syed Z. Arshad
National ICT Australia (NICTA)
Sydney, NSW
Australia

Dan Conway
National ICT Australia (NICTA)
Sydney, NSW
Australia

Jianlong Zhou
National ICT Australia (NICTA)
Sydney, NSW
Australia

Kun Yu
National ICT Australia (NICTA)
Sydney, NSW
Australia

Ahmad Khawaji
National ICT Australia (NICTA)
Sydney, NSW
Australia

ISSN 1571-5035 ISSN 978-3-319-31700-7 (electronic)
Human–Computer Interaction Series
ISBN 978-3-319-31698-7 ISBN 978-3-319-31700-7 (eBook)
DOI 10.1007/978-3-319-31700-7

Library of Congress Control Number: 2016937498

Printed on acid-free paper

This Springer imprint is published by Springer Nature
The registered company is Springer International Publishing AG Switzerland

Preface

The body of knowledge presented in this book is timely. As we enter the age of data, human capacities are increasingly the critical factor in determining the output of human-machine interactions. This has important implications not only in high reliability operations such as aviation, command and control and complicated industrial and commercial applications, but also in everyday device usage by the general population. In order to achieve optimal 'human-in-the-loop' system behaviour, it is therefore increasingly important to understand and adapt to the constraints of human cognitive abilities.

The research presented herein is the result of a unique combination of people and circumstances. A research group within NICTA had been carrying out research into interface technologies and for some years, producing both basic research and engineered applications. The team had a track record in the investigation of speech, linguistic features, gestures and then later physiological signals such as Galvanic Skin Response (GSR), eye-based signals, pen gestures and electroencephalography (EEG), all in the context of developing cutting-edge user interfaces. These investigations then evolved into a research focus on the advantages of multimodal interfaces and how cognitive architectures, when linked with technology, can be leveraged to provide more efficient and responsive Human-Computer Interaction (HCI).

Sometime in late 2004, during a site visit to a traffic emergency response facility, a manager asked me the following question: 'All my operators typically have multiple issues open that they are working on at any one time. When something dramatic and urgent occurs that requires immediate attention – who should I give the additional issue to?'

Armed with its previous experience, our team thus focused its attention on investigating and operationalizing the construct of cognitive load and its effect on user performance with the aim of providing robust, real-time quantification tools and methods. This work was then extended into the design of adaptive systems (both human and machine) that are able to compensate for the intersection of task demands and a user's cognitive capacities. Answering these questions has involved a 10 year research campaign involving 13 staff, 16 PhD students, numerous other researchers and along the way produced over 70 papers, four patent families (applied and granted in countries including the USA, Canada and Australia), finally culminating in this book.

A fundamental conceptual principle that emerged from this body of research is that the ability to observe and quantify the end-user through sophisticated data collection techniques enables systems to adapt to constantly changing work environments. This is achieved by learning about the intersection of multiple sources of variance such as the different characteristics of each feature evaluated for each task component and in each context. Furthermore, the data analytics expertise within the group allowed an unprecedented level of analysis of the complicated signals generated by both human behaviour and physiology.

It should be noted that the team that has carried out much of the research presented here is situated firmly within a data-driven approach and is at the vanguard of machine learning and algorithmic technologies, which manifests itself within this work in both methods and applications. Many analyses presented here are machine learning based, but are discussed in general terms, thus being accessible to those not familiar with the methods themselves. The text is aimed at researchers and engineers generally, including undergraduates in HCI, psychology and related fields. There is also an emphasis on practical problems and applied research, as a result of the ongoing mandate of our team to grapple with pressing issues in the real world.

This book reviews the current state of play in the literature as well as presents our own research into multimodal cognitive load measurement focusing on non-intrusive physiological and behavioural modalities, such as signals derived from the eye, GSR, speech, language, pen input, mouse movement as well as multimodal approaches. Factors which affect cognitive load measurement such as stress, trust and environmental factors such as screen illumination are also discussed. Furthermore, dynamic workload adjustment and real-time cognitive load measurement with data streaming are presented. Finally, typical application examples of cognitive load measurement are reviewed to show the feasibility and applicability of multimodal cognitive load measurement to situated applications. This is the first book to systematically introduce various computational methods for automatic and real-time cognitive load measurement and by doing so moves the practical application of cognitive load measurement from the domain of the computer scientist and psychologist to more general end-users ready for widespread implementation.

Whilst there is more work to be done in the field, the overwhelming message that should be taken away from this book is that many of the approaches outlined herein have now been validated and that useful and responsive measurement of human cognitive load is both achievable and an important factor in bringing about a more perfect union with our machines.

Sydney, NSW, Australia Dr. Fang Chen
January 2016

Acknowledgements

There are many people who have contributed to the multimodal cognitive load measurement over the past 10 years in our group at National ICT Australia (NICTA). Our first thank is given to all staff and students who had worked on cognitive load measurement during the past years at NICTA, they are Julien Epps, Natalie Ruiz, Bo Yin, Eric Choi, Yu Shi, M. Asif Khawaja, Nargess Nourbakhsh, Pega Zarjam, Siyuan Chen, Sazzad Hussain, Tet Fei Yap, Phu Ngoc Le, Ronnie Taib, Benjamin Itzstein, Nicholas Cummins, Ling Luo, Ju Young Jung and many others. Without them, this book would not have been written.

We also thank many volunteer participants from universities such as UNSW, USYD and various NICTA groups who contributed their precious time for many different cognitive load measurement experiments conducted at NICTA. These people, in many instances, are students who are busy with their study and staff who have time pressure for their deliverables gave freely of their time and experiences thereby enabling us, as cognitive load researchers, to research this fascinating topic. Their great contribution makes our cognitive load measurement experiment successful.

We acknowledge our collaborators from around the world who have had various discussions and comments on our work.

Lastly, and most importantly, we would like to thank NICTA providing all kinds of support in the past years to make our research and experiments run smoothly and successfully.

Contents

Part I
Preliminaries

Chapter 1
Introduction

Cognitive load is an increasingly important determinant of the performance of human-machine systems. As the power, complexity and ubiquity of both everyday and specialised computing solutions increase, the limiting factor of such systems performance is shifting from the machine to the human. This dynamic is further exacerbated by the need for human decision making in environments characterized by the need to process and integrate an unprecedented amount of information. Furthermore, the rapid pace of industrial, commercial, and indeed every-day life, frequently requires high levels of user performance under time-limited conditions and often in sub-optimal contexts characterised by stress and competing demands for attention. As such, the need to understand and design for the particular characteristics of the user has never been stronger.

Cognitive load has been shown to have a fundamental, direct, albeit complicated relationship with human performance in HCI (Human-Computer Interaction) contexts. In general terms humans have been shown to be able to increase cognitive effort to match increasing task demands up until they reach the limit of their mental capacities. Beyond this point, as task demands increase, human performance declines and is characterized by an increase in errors as well as secondary effects such as stress and negative affect, as shown in Fig. 1.1. Performance is also influenced by additional factors such as individual differences in capacity and expertise, as well as contextual variables such as stress, motivation, affect and the environment generally.

The relationship between task demands, mental capacity and performance is explained through the construct which is central to this book, namely Cognitive Load (CL).

F. Chen et al., *Robust Multimodal Cognitive Load Measurement*, Human–Computer Interaction Series, DOI 10.1007/978-3-319-31700-7_1

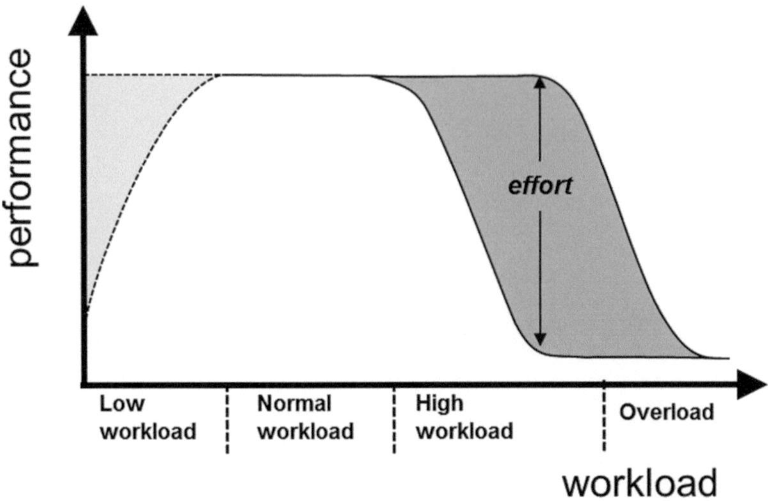

Fig. 1.1 The relationship between task demands, performance, and workload (Adapted from [1], the use of the source figure was permitted by Veltman and Jansen)

1.1 What Is Cognitive Load

Whilst we provide an account of the latest research findings regarding the construct in Chap. 2, and explore theoretical aspects in Chap. 3, it may be useful to outline initially, in general terms, our conceptualization of cognitive load. The construct of cognitive load is based on models of human working memory which state that we have limited capacity to process information. A foundational finding in this respect was Miller's work [2] that suggested that we can only hold 7 (plus or minus two) "chunks" of information in working memory at one time and our information processing ability is therefore restricted to these entities. Cognitive load is therefore a variable that attempts to quantify the extent of demands placed by a task on the mental resources we have at our disposal to process information. Whilst there have been numerous, both overlapping and sometimes divergent definitions of this construct in the literature, this book generally adopts the following, reasonably consensual, definition: "*…it is a multi-dimensional construct representing the load imposed on the working memory during performance of a cognitive task*" [3, 4].

Cognitive load is highly dynamic, and can change from second to second, even within the execution of a given task. Investigations into the subject were initially dependent on subjective experience and were operationalised around subjects rating task difficulty levels or the amount of mental effort required to complete a task.

1.2 Background

Attempts to assess cognitive load have been in large part driven by the requirements of work environments where human errors are highly undesirable. High accuracy and time-sensitive environments such as emergency response, aviation, incident management as well as military command and control instances have a low human fault tolerance and as such have driven recent research efforts into the assessment of cognitive load. An eventual agenda of this enquiry is to be able to both predict user performance in relation to task demands, and then ultimately adapt workflows to compensate for human limitations, and thereby avoiding performance failures owing to task demands that exceed human mental capacities. The applications of cognitive load measurement are, however, not limited to these error-critical environments, but also hold great promise for the design of systems and software designed for everyday use by the general population.

Although numerous methods have been proposed and investigated with the aim of measuring cognitive load, it remains a difficult construct to assess – in particular because of its multi-dimensional nature. A particular body of research, that we will refer to as Cognitive Load Theory (CLT) [5], has focused on human learning and proposes a model that decomposes the construct into three component factors. Intrinsic load is associated with the nature of the learning material itself and is essentially unavoidable. Germane cognitive load refers to the cognitive resources dedicated to constructing new schema in long-term memory – i.e.: the actual processing required to perform the task itself. And extraneous load refers to the representational complexity – that is, complexity that varies depending on the way the task is presented, with the assumption that extraneous load can be minimized in order to allow more processing for intrinsic load.

For example, in a traffic management scenario, an example task may be to find the exact location of an accident. The equipment, tools and applications the human employs to complete the task, for example, a paper-based directory, a Geographical Information System (GIS) or electronic maps, or even street monitoring cameras, each contribute to extraneous load. Understanding of the location itself can be considered intrinsic load, and germane load the integration with other task demands.

A great variety of cognitive load measuring techniques, ranging from simple approaches such as questionnaires to highly involved procedures such as functional braining imaging techniques, have been used to study cognitive load [6–8]. Generally, these measures can be divided into the following categories: subjective ratings, performance measures, behavioural measures and physiological measures [9–11]. Subjective ratings are generally used as ground truths in cognitive load experiments but have the disadvantage of taking place after the event. Performance measures offer a good temporal relationship to task demands, but are generally insensitive to variations in load up until they approach or exceed cognitive capacities. Behavioural and physiological measures hold promise in providing a more direct and immediate method to assess cognitive load but establishing a direct relationship between such measures and a user's cognitive state can be problematic. On the other hand these measures are less likely to interfere with user's performance on primary tasks and are often discussed in terms of being non-intrusive.

1.3 Multimodal Cognitive Load Measurement

Attempts to measure cognitive load via physiological measures such as GSR, eye activity, respiration and heart rate have become increasingly sophisticated and their granularity and high degree of responsiveness hold much promise, but, in nearly all cases these measures suffer from limitations from both confounding factors and noise. As a result, this book spends significant time detailing approaches where multiple human-derived signals are analysed (see Fig. 1.2), which we refer to as a multi-modal approach to cognitive load measurement. We will go on to show that multimodal approaches can overcome the limitations of individual signals and provide more robust representations of cognitive load than can be derived from any one data source. As an example, in an arithmetic task, eye blink rate and GSR response were shown to generate higher accuracy of cognitive load measurement when the signals were fused, than each measure alone.

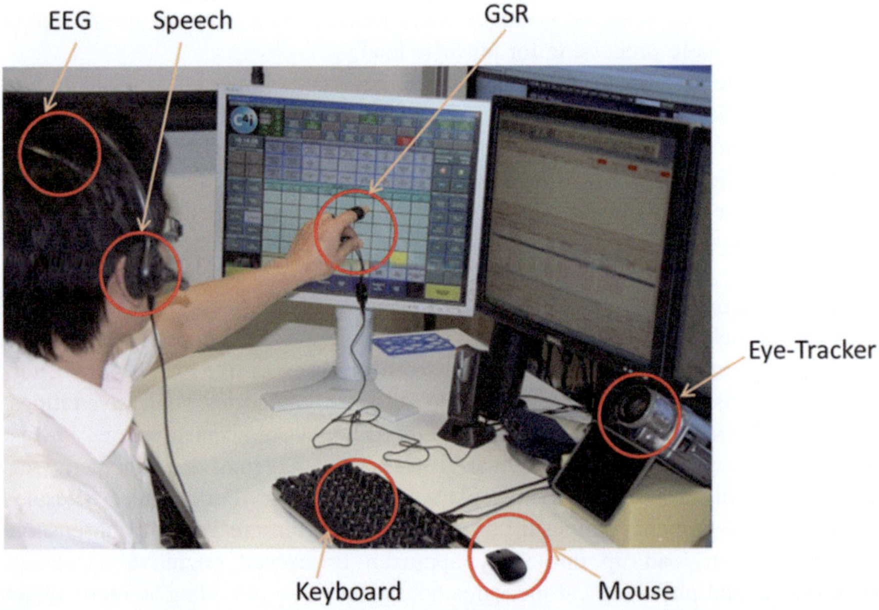

Fig. 1.2 Example modalities used for cognitive load measurement

Multimodal approaches have also been shown to be increasingly robust to non-work related measures such as luminance changes and other environmental factors. Furthermore they offer utility in that each individual measure may only be sensitive to a certain range of load. One measure may show excellent differentiation between various lower levels of load, but then exhibit ceiling effects where it cannot provide disambiguation at higher load levels. Multimodal approaches can be used to overcome these limitations and extract a more information than any one index.

Multimodal approaches have also been supported by an increasing body of evidence derived from neuroimaging methods such as positron emission topography and functional magnetic resonance imaging. These have been used to identify separate locations in the brain for the verbal/auditory, imagery/spatial and executive functions of working memory [12, 13].

Figure 1.3 gives an overview of multimodal cognitive load measures. As shown in this figure, cognitive load can be measured with four kinds of methods as mentioned previously: subjective ratings, physiological signal based measures, behavioral signal based measures, and task performance based measures. These measures can also be categorized into data-driven measures and knowledge-based measures. These different modalities can be fused for multimodal cognitive load measurement.

Unlike the cognitive load measurement approaches prevalent in the psychology of learning, which mainly rely on subjective ratings for cognitive load investigation, this book evaluates cognitive load from the perspective of human responses, i.e., our methods are on a more *data-driven* basis. We focus on behavioural and

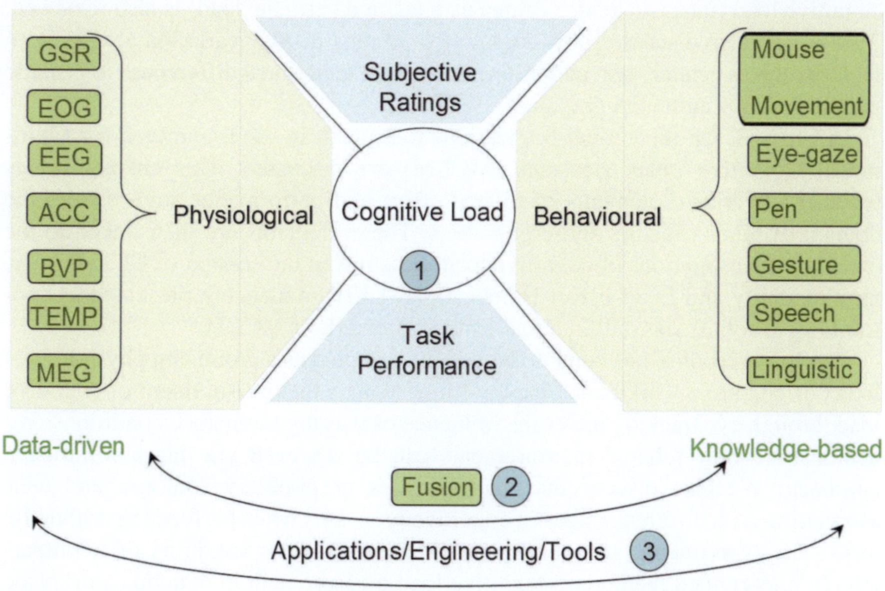

Fig. 1.3 An overview of cognitive load measures (GSR: Galvanic Skin Response, EOG: Electrooculography, EEG: Electroencephalogram, ACC: Anterior Cingulate Cortex signal, BVP: Blood Volume Pulse, TEMP: Temperature, MEG: Magnetoencephalography)

physiological signals, which include signals derived the eye, GSR, speech, linguistic features, pen input, mouse movement as well as multimodality fusions. These signals can be recorded and analysed (sometimes in real-time) with the aid of data analytics tools such as machine learning. Factors which affect cognitive load measurement such as stress, trust, and environmental factors such as emotion are also extensively investigated in this book.

The eventual aim of these approaches is to be able to develop adaptive systems, where task parameters are modulated by signals derived from the quantification of cognitive load, with the aim of keeping the difficulty within the range of cognitive processing available to the human.

1.4 Structure of the Book

This book is organized as shown in Fig. 1.4.

In recent years, there has been an increased amount of research into cognitive load from various disciplinary backgrounds. Chapter 2 provides an overview of the latest developments in this body of work. Firstly, different approaches to the problem of measuring cognitive load are identified. The related work of each measurement approach is then discussed in more detail. These approaches include subjective (self-report) measures, physiological measures, performance measures, and behavioural measures. The related work on measuring different types of cognitive load (intrinsic load, extraneous load, and germane load) is also reviewed. This review also discusses how cognitive load may display variation according to factors such as gender, age, and information representational differences (e.g. static graphics versus animated graphics).

In Chap. 3, the theoretical foundations that result in the framework for Multimodal Cognitive Load Measures (MCLM) are discussed. This chapter brings together the critical elements of mental effort and performance studies from the domain of human factors and ergonomics. These elements are then linked to the extensive investigations of cognitive load measures in the context of CLT. Finally, multimodality and Load Effort Homeostasis (LEH) models are presented as contributing to robustness within a real-time framework of MCLM.

Pupillary response has come to be widely accepted as a promising physiological index of cognitive workload. Chapter 4 investigates the measurement of cognitive load through eye tracking under the influence of varying luminance conditions. We demonstrate how reliable measurements can be achieved via this non-intrusive approach. We also discuss the characteristics of pupillary response and their association with different stages of cognitive processes when performing arithmetic tasks. The experimental results presented demonstrate the feasibility of a comparatively fine-grained method of cognitive load measurement in dynamic workplace environments.

Chapter 5 focuses on the use of GSR for cognitive load measurement. GSR is a measure of conductivity of human skin, and provides an indication of changes

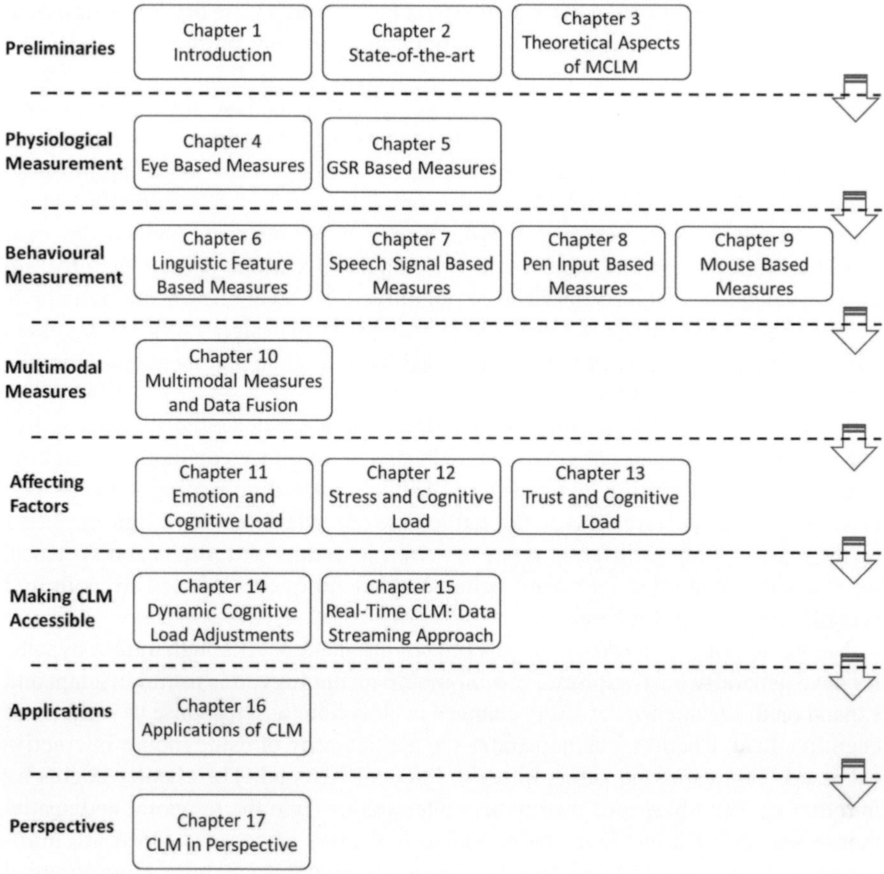

Fig. 1.4 Organization of the book

within the human sympathetic nervous system. It has recently attracted researchers' attention as a prospective physiological indicator of both cognitive load and emotional responses. We investigate temporal and spectral features of GSR data captured from two different experiments, one utilising a text reading tasks and the other arithmetic tasks, each imposing multiple cognitive load levels. Blink and GSR features and combinations thereof are then used for cognitive load classification. The results show that accumulative GSR, the power spectrum of GSR, blink number and blink rate are significantly distinctive and have reasonable accuracies in both two- and four-class classification of cognitive load using Support Vector Machines and Naïve Bayes classifiers.

Words, phrases and sentences are basic linguistic components used to exchange information between people. Recent investigations have revealed that people exhibit different patterns of language expression when experiencing differing cognitive load levels. Chapter 6 discusses methods for cognitive load examination via language and shows that many features, some of which were originally designed to examine language complexity for learning analytics or comprehension, can be successfully applied to the cognitive load research. The linguistic approach to measure cognitive load can also be used as a post-hoc analysis technique for user interface evaluation and interaction design improvement using speech transcripts.

Speech has also been found to be affected by cognitive load, including prosody, spectrum and vocal tract related parameters. Chapter 7 reviews cognitive load measurement methods via speech with typical experiments to induce different cognitive load levels via speech being introduced. Various speech features are then investigated, and a comparison of cognitive load level classification methods is conducted. Additionally, cognitive load-aware system design issues are discussed with an emphasis on real-time cognitive load measurement and its potential implications for system usability.

Writing with a pen, as a complex form of interaction, often entails intensive user attention and associated cognitive load. Chapter 8 introduces methods to examine cognitive load via writing and pen-based features, including writing velocity, pen pressure, writing gestures and additional features derived from these signals. Based on the examination of different types of script, including text, digits and sketches, we show that cognitive load with behavioural features are affected by both text content and writing direction.

Mouse activity is an important part of user interaction. Although mouse dynamics have generally been explored as a biometric technology, it is useful to adapt and expand on this usage for detecting changes in behaviour as a response to variance in cognitive load. Chapter 9 demonstrates the applicability of using mouse interactivity based features as indicators of a user's cognitive load. The chapter begins by introducing some basics of user mouse interaction. Then the temporal and spatial mouse features that are found to be viable indicators of cognitive load are introduced. The possibility of incorporating mouse interactivity features in multimodal cognitive load measures is also assessed in this chapter.

Previous chapters have focused on utilising features from various single modalities that allow us to differentiate between cognitive load levels. However, expanding analysis beyond single-modality indices, to a multi-modal approach, may offer a way in which to increase measurement robustness. Chapter 10 presents a model for multimodal cognitive load. The features extracted from speech, pen input and GSR in a user study are fused using the AdaBoost algorithm to demonstrate the methods advantages.

In practice, various confounding factors unrelated to cognitive load, including changes of luminance conditions and emotional arousal are likely to degrade the accuracy of cognitive load measurement. Chapter 11 investigates pupillary response and GSR as a cognitive load measure under the influence of such confounding factors. A video-based eye tracker is used to record pupillary response and GSR is recorded during arithmetic tasks under both luminance and emotional changes. The mean-difference feature and its extension (Haar-like features) are used to characterise physiological responses of cognitive load under these context effects. Boosting based feature selection and classification are employed that successfully classify cognitive load even under the influence of those noisy factors.

Besides luminance conditions and the emotional arousal we investigate in Chap. 11, other factors such as the presence of stress may affect physiological measurement in ways that confound reliable detection of cognitive load. Chapter 12 investigates the effect of stress on cognitive load measurement using GSR as a physiological index of cognitive load. The experiment utilises feelings of lack of control, task failure and social-evaluation to induce stress. The experiment demonstrates that mean GSR can be used as an index of cognitive load during task execution, but this relationship is obfuscated when test subjects experience fluctuating levels of stress. Alternate analysis methods are then presented that show how this confounding factor can be overcome.

Trust has been found to be an important factor in determining aspects of human behavior with many types of systems, especially complex, high-risk domains such as aviation and military command and control. Chapter 13 investigates the relationship between trust perception and cognitive load. An experimental platform is designed and employed to collect multimodal data and different types of analyses are conducted.

By monitoring a user's state and adapting task difficulty levels, dynamic cognitive load systems promise to be able to improve users performance and help users maximize their capacity for productive work. Chapter 14 presents a cognitive load adaptation model that dynamically adjusts cognitive load during human-machine interaction, in order to keep the task demands at an appropriate level. Physiological signals such as GSR are collected to evaluate cognitive load in real-time and the task difficulty levels are adjusted in real-time to better fit the user.

Building on the concepts presented previously, in Chap. 15 we discuss how the efficacy of intelligent user interfaces would be greatly enhanced if a user's cognitive load could be sensed in real time and adjustments made accordingly. Different data streams derived from the user can individually (or collectively) be processed to detect sudden *shifts* or gradual *drifts* in behavior. We present a reformulation of the problem of cognitive load change detection as the problem of shift/drift detection in data streams. The chapter extends the multimodal behavioral model to include mouse interactivity streams and discusses a modified sliding window implementation. In detailing an experiment utlising several variations of this model, the technical feasibility of learning from streams sheds light on the challenges presented by real time cognitive load measurement.

In Chap. 16 we present some typical application examples of cognitive load assessment and demonstrate the feasibility and applicability of multimodal cognitive load measurement approaches in various applications and instances of HCI. In this chapter, we firstly show how cognitive load measurement can be used in adaptive user interface design in order to improve user interaction efficiency. We then discuss how cognitive load theory can be used in emergency management where people's cognitive fatigue and mental overload can result in catastrophic outcomes. Cognitive load theory is also widely used in education and training to improve learning efficiency. It can also be used in critical situations such as monitoring the states of car drivers, pilots and aircraft maintenance teams. These application examples are reviewed in details.

Chapter 17 provides a summary of all the work presented and future perspectives of cognitive load measurement are discussed.

References

1. J. Veltman, C. Jansen, *The Role of Operator State Assessment in Adaptive Automation*. TNO-DV3 2005 A245 (TNO Defence, Security and Safety, Soesterbery, 2006)
2. G.A. Miller, The magical number seven, plus or minus two: Some limits on our capacity for processing information. Psychol. Rev. **63**(2), 81–97 (1956)
3. F.G.W.C. Paas, J.J.G.V. Merriënboer, Instructional control of cognitive load in the training of complex cognitive tasks. Educ. Psychol. Rev. **6**(4), 351–371 (1994)
4. F. Paas, J.E. Tuovinen, H. Tabbers, P.W.M. Van Gerven, Cognitive load measurement as a means to advance cognitive load theory. Educ. Psychol. **38**(1), 63–71 (2003)
5. J. Sweller, P. Ayres, S. Kalyuga, *Cognitive Load Theory* (Springer, New York, 2011)
6. R. Gingell, *Review of Workload Measurement, Analysis and Interpretation Methods*. European Organisation for the Safety of Air Navigation (2003)
7. M.A. Just, P.A. Carpenter, A. Miyake, Neuroindices of cognitive workload: Neuroimaging, pupillometric and event-related potential studies of brain work. Theor. Issues. Ergonomics Sci. **4**, 56–88 (2003)
8. W.W. Wierwille, F.T. Eggemeier, Recommendations for mental workload measurement in a test and evaluation environment. Hum. Factors: J. Hum. Factors. Ergonomics. Soc. **35**(2), 263–281 (1993)
9. S. Hart, L. Staveland, Development of the NASA-TLX (Task Load Index): Results of empirical and theoretical research, in *Human Mental Workload*, ed. by P.A. Hancock, N. Meshkati (North Holland Press, Amsterdam, 1988)
10. R. O'Donnell, F. Eggemeier, Workload assessment methodology, in *Handbook of Perception and Human Performance. Cognitive Processes and Performance, vol. 2*, ed. by K. Boff, L. Kaufman, J. Thomas (Wiley, New York, 1986), pp. 42–1–42–49
11. G.F. Wilson, R.E. Schlegel, *Operator Functional State Assessment*. NATO RTO Publication RTO-TR-HF M-104 (NATO Research and Technology Organization, Neuilly sur Seine, France, 2004)
12. E.E. Smith, J. Jonides, R.A. Koeppe, Dissociating verbal and spatial working memory using PET. Cereb. Cortex **6**(1), 11–20 (1996)
13. E. Awh, J. Jonides, E.E. Smith, E.H. Schumacher, R.A. Koeppe, S. Katz, Dissociation of storage and rehearsal in verbal working memory: Evidence from positron emission tomography. Psychol. Sci. **7**(1), 25–31 (1996)

Chapter 2
The State-of-The-Art

Recent years have witnessed booming growth in the research of cognitive load understanding and measurement [1]. Extensive research has been carried out into cognitive load theory, cognitive load measurement, applications of cognitive load theory, and the factors that may affect cognitive load. This chapter surveys the state-of-the-art of cognitive load measurement and gives an overview of current research in cognitive load measurement. The highlights include:

- Introduction of three types of cognitive load;
- Overview of four categories of cognitive load measurement methods, namely: subjective measures, performance measures, physiological measures, and behavioral measures;
- Measurement of different types of cognitive load;
- Factors that may cause differences in cognitive load, such as gender, age, graphics presentation.

2.1 Working Memory and Cognitive Load

It is well-established that the two main limitations of working memory resources are its capacity and duration [2]. According to Baddeley's model [2], working memory has separate processors for visual and verbal information. Only a limited number of items, or "chunks", can be held in working memory at any one time and then only for limited duration [3]. These limitations are never more evident than when users undertake complex tasks, or when in the process of learning – when extremely high demands are placed on working memory. The construct of cognitive load refers to the working memory demand induced by a complex task in a particular instance where novel information or novel

F. Chen et al., *Robust Multimodal Cognitive Load Measurement*, Human–Computer Interaction Series, DOI 10.1007/978-3-319-31700-7_2

Fig. 2.1 Three types of cognitive load

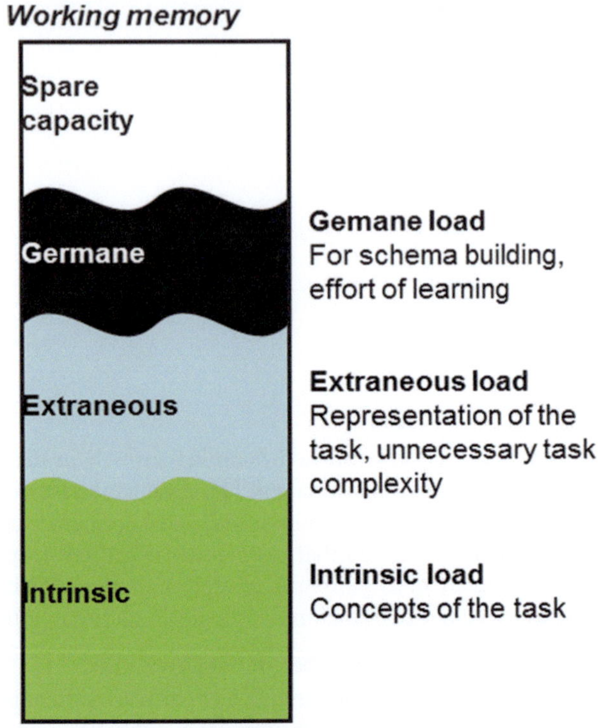

processing is required [4]. Any single task can induce different levels of mental effort or cognitive load from one user to another, or as a user gains expertise. This discrepancy in the mental demand from person to person could be due to a number of reasons, for example: level of domain expertise or prior knowledge, interface familiarity, the user's age, or any mental or physical impediments. A task that may cause high load in one user may not necessarily do so in a more experienced user.

The cognitive load construct comprises of at least three separate load sources: *intrinsic load, extraneous load,* and *germane load* [4, 5] (see Fig. 2.1). Intrinsic load is associated with the inherent challenge or level of difficulty of the material being processed. Extraneous load refers to the representational complexity – that is, complexity that varies depending on the way the task is presented. Germane cognitive load refers to the cognitive resources dedicated to constructing new schema in long-term memory. All these types of load combine to form the overall experience of cognitive load. Situations that induce high levels of cognitive load can impede learning and efficient performance on designated tasks [4, 5].

The ability to determine exactly when a user is being cognitively loaded beyond a level that they are able to manage could enable the system to adapt its interaction strategy intelligently. For example, the system could attempt to reduce the cognitive load experienced by the human – particularly in terms of extraneous load – such that optimal performance is facilitated. A number of methods have been used, both in HCI and other domains, to estimate the level of cognitive load experienced. There are four main methods comprising the current state of the art [1]:

- **Subjective (self-report) measures**, where users rank their experienced level of load on single or multiple rating scales [6];
- **Performance measures**, such as task completion time, speed or correctness, critical errors and false starts [7–9], as well as performance on a secondary tasks [10];
- **Physiological measures**, such as galvanic skin response, and heart rate [11];
- **Behavioral measures**, which observe feature patterns of interactive behavior, such as linguistic or dialogue patterns [12], and even text input events and mouse-click events [13].

However, while most of these types of measures are suitable for research purposes, many are not feasible for widespread deployment in interactive intelligent systems.

2.2 Subjective Measures

Traditionally, the most consistent results for cognitive load measurement have been achieved through subjective measures [7]. These measures ask users to describe in fine details and reflect each user's perception of cognitive load by means of introspection: the user is required to perform a self-assessment of their mental demand by answering a set of assessment questions immediately after the task.

There are two types of scales of subjective ratings:

- **Unidimensional scales**, which measure overall cognitive load like Subjective Cognitive Load Measurement scale [14].
- **Multidimensional scales**, which focus on the different components of load [16].

According to CLT, unidimensional scales are criticized because they cannot capture the multifactorial nature of cognitive load. It is also nevertheless recognized that a single item of difficulty is a good indicator of overall cognitive load [15]. On the other hand, a multidimensional scale such as the NASA Task Load Index (NASA-TLX) [16] gives a broader evaluation of cognitive load. It is based on six dimensions (performance, mental effort, frustration, task demand, physical demand, temporal demand). Cierniak et al. [17] proposed a three-item scale which includes a difficulty rating for the learning content, difficulty of the material and concentration during learning. As a result, it assesses intrinsic, extraneous, and germane load respectively. More recently, Leppink et al. [18] have developed a ten-item subjective cognitive load scale with students attending lectures on statistics. In their approach, they used multiple items for each type of cognitive load in order to increase the accuracy: (1) intrinsic load was evaluated with three items referring to the complexity of the material; (2) extraneous load was measured with three items related to the negative characteristics of information providing during the classes; and (3) germane load was evaluated by the mean of four items that dealt with the contribution of explanations and instructions to the learning. Despite these promising approaches, there are some limitations inherent to self-reported measures. For instance, subjective scales are often administered after the learning task and, in this way, are not easily able to capture variations in load over time [14]. Such an approach is also impractical in real, day to day situations because the questionnaires not only interrupt task flow but also add more tasks to an already potentially overloaded user [1].

2.3 Performance Measures

Performance-based approaches provide measures that can reflect variations encountered during the task. The basic assumption for performance measures is that learning is hindered when working memory capacity is overloaded, and therefore a drop in performance will be the result of an increase in overall cognitive load [14]. One of the most widely used techniques is the dual-task paradigm, in which the performance is evaluated in a secondary task performed in parallel to assess the cognitive load devoted to the main task. The hypothetical relationship between performance and workload as discussed by O'Donnell and Eggemeier [7], is composed of three regions, Low, Medium and High. The authors claim that primary task measures of workload cannot be used to reflect mental workload in region Low, because this region is characterized as indicating "adequate task performance" on behalf of the subject [7]. However, in many real world tasks, what constitutes "adequate task performance" is analogous to a band of acceptable outcomes rather than a single correct response, and subtle differences may occur between different solution alternatives which may not be reflected in the overall performance measures used. Certain features of the behavioral responses have the potential to differentiate between these solution outcomes by identifying compensatory behaviors [1]. In much the same way, performance measures cannot measure spare capacity [19], when a user still has plenty of cognitive resources to deploy.

In the "Medium" region, both primary and secondary task performance measures can be used to reflect workload as performance decreases. Dual-task approaches have been incorporated in several studies to measure subjects' performance in controlled conditions [7]. While secondary task performance can provide a measure of remaining resources not being used by the primary task [20, 21], it is not feasible for humans to complete dual tasks "in the wild", and hence these cannot be adopted for widespread use. In real world tasks, performance measures from the primary task can be extremely difficult to calculate on the fly, if at all. In the case of Transport Management Centers, senior staff will often conduct reviews of incident handling to debrief humans and qualitatively rate their performance. In this application, access to automatic cognitive load estimates around every hour or so would be considered a dramatic improvement for review purposes. Assessment on per minute basis or so would be considered "real time" and could be used to directly affect moment-to-moment human resource allocation.

Performance measures tend to remain stable as load increases in the "Medium" region, particularly when the human exerts a greater amount of mental effort, as noted by [7]. This is addressed more specifically by Veltman and Jansen [22] who proposed a range within which compensatory efforts may have an effect. Figure 1.1 illustrates this concept – the subject still achieves a high level of performance within the region labeled "effort", depending on the degree of effort exerted. Exposure to high cognitive load culminates in a higher likelihood of errors [23–25] and compensatory efforts can only be maintained for a time after which a subject then fatigues and their performance begins to decline [24]. At the overload stage, compensatory efforts no longer make a difference – it is too late for the system to react appropriately to ease the human's load and both the system and the user must engage in (typically costly) recovery strategies.

The approaches described thus far have the disadvantage of being physically or psychologically intrusive. Likewise, many of them are also post-hoc and hence not conducive to their implementation as real-time adaptive behaviors and strategies by an interactive intelligent system or interface. Performance measures can also depend on the human completing the task, which may not always be possible in high load situations – for example, the human may be stuck on one or two steps of the overall task for a relatively long period of time where no valid task-based performance assessment can be calculated.

Similarly, performance measures – which are often defined as measures that reflect the accuracy and correctness of a user's response and are directly relevant to the outcome of the task – are often calculated after the fact, if they can be assessed objectively at all. In some complex domains, measures based on performance outcomes are impossible to access in real-time such that the system would be able to act on the information in a timely manner. For example, the spontaneous nature of crisis management and other control room situations means the user's performance in this sense is very difficult to rate, even during debriefing, and unique to almost every situation. The actions taken can vary widely from human to human, both in order and content, while still being equally effective in achieving the task goals and solving the problem to an adequate level of performance. In some cases, performance cannot be calculated automatically at all [1].

2.4 Physiological Measures

Research has found that changes in various physiological states are closely related to changes in cognitive load [1]. One major advantage of physiological measures is the continuous availability of bodily data, allowing cognitive load to be measured at a high rate and with a high degree of sensitivity. This research has also shown that different brain regions control different cognitive and mental activities, for instance, the dorsolateral prefrontal cortex (DLPFC) is associated with intrinsic load in cognitive activities [26]. The functional imaging and clinical evidence indicates that a remarkably consistent network of brain regions is involved in cognitive activities [27]. Moll et al. [28] reviewed evidence on brain regions identified during functional imaging of cognition related activities irrespective of task constraints. It was demonstrated that the investigation of mechanisms of cognition–emotion interaction and of the neural bases is critical for understanding of the human mind. Van Gog et al. [29] applied an interdisciplinary approach combining evolutionary biological theory and neuroscience within a cognitive load framework [30] to explain human's behaviour in observational learning. Because of the close relationship between cognitive load and neural systems, human neurophysiological signals are seen as promising avenues to measure cognitive load [31].

The physiological approach for cognitive load measurement is based on the assumption that any changes in the human cognitive functioning are reflected in the human physiology [31]. The measures that have been used in the literature to show some relationship between subjects' cognitive load and their physiological behavior include:

- Heart rate and heart rate variability [32–34];
- Brain activity (e.g. changes in oxygenation and blood volume, electrocardiography (ECG), electroencephalography (EEG)) [35, 36];
- Galvanic Skin Response (GSR) or skin conductance [37, 38];
- Eye activity (e.g. blink rate, pupillary dilation) [39–44].

This book focuses on pupillary dilation (Chap. 4) and GSR (Chap. 5) to show how physiological features are used to index cognitive load.

Chen and Epps [43] investigated three types of eye-based approaches for Cognitive Load Measurement (CLM): pupillary response, blinking, and eye movement (fixations and saccades). Eye-based features were investigated in the presence of emotion interference by affective image. An experiment with arithmetic-based tasks demonstrated that arousal effects were dominated by cognitive load during task execution. A set of features such as zero crossing count of pupil size, features from cumulative blink/fixation/saccade numbers, eye features from task stimuli onset to the first saccade were all proposed for cognitive load level prediction. The performance of cognitive load level prediction was found to be close to that of a reaction time measure, showing the feasibility of eye activity features for near-real time CLM. Hossain and Yeasin [45] introduced a Hilbert transform analytic phase based method to compute temporal patterns from pupillary responses. Arithmetic multiplication tasks were used in the study to collect the data. Analysis shows that sharp changes in human response signals from tasks that performed by users incorrectly may be attributed to a cognitive overload. It was also observed that a sharp change and continuation of the ramp in Hilbert unwrapped phase relates to the cognitive dissonance. Ledger's [46] experiments showed that blink rate decreases as cognitive load increases. Das et al. [47] investigated the use of unsupervised learning approach to measure cognitive load with EEG signals. The results indicated that the unsupervised approach is comparable and sometimes better than supervised (e.g. Support Vector Machine (SVM)) method. Further, in the unsupervised domain, the Component based Fuzzy c-Means (CFCM) outperforms the traditional Fuzzy c-Means (FCM) in terms of the CL measurement accuracy.

Changes in the physiological data occur with the level of stimulation experienced by the person and can represent various levels of mental processing. The data collected from body functions are useful as they are continuous and allow the signal to be measured at a high rate and in fine detail. However, physiological measures require users to wear a lot of cumbersome equipment, e.g. EEG headsets that not only interfere with users' task, but are prohibitive in cost and implementation. Additionally, the large amounts of physiological data that need to be collected and the expertise needed to interpret those signals render many types of physiological signals unsuitable for common interactive intelligent systems [11]. While they can be very sensitive to cognitive activity, the above issues in combination with the degree of variability of physiological signals, due to external factors such as temperature and movement, means they may have limited suitability for everyday environments [11].

Joseph [48] compared different CLM methods of self-report and physiological measures based on eye-tracking and EEG. The comparison found that physiological features from eye-tracking and EEG were sensitive to tasks that varied in the level of intrinsic load. The self-report measures performed similarly when the difference in intrinsic load of tasks was the most varied.

Table 2.1 lists some of the recent research work using physiological measures for cognitive load evaluation in various cognitive tasks.

2.5 Behavioral Measures

Response-based behavioral features are defined as those that can be extracted from any user activity that is predominantly related to deliberate/voluntary task completion. Examples include:

- Eye-gaze tracking;
- Mouse pointing and clicking;
- Keyboard usage;
- Gait patterns;

Table 2.1 Cognitive task – physiological measure matrix

Task	BF	BI	BD	FF	FD	SD	SS	PD	PC	PS	IC	ST	HM
Air traffic control task [49]	•		•			•		•					
Air traffic control task [50]	•					•							
Air traffic control task [51]						•	•						
Auditory two-back task [52]	•							•					•
Cart driving and stationary bike exercise [53]								•					
Cognitive task and visual search task [54]	•							•					
Combat management task [55]					•	•	•	•					
Continuous memory task [56]												•	
Division task and Sternberg memory search [57]										•			
Driving task and auditory addition task [58]	•		•	•	•			•					
Driving task and secondary task [59]											•		
Driving task and spoken task [60]								•					
Driving task and verbal/spatial-imagery task [61]								•					
Document editing, email classification, route planning [62]									•				
Flight task and memory task [63]		•	•										
Flight task with visual/instrument flight rule [64]	•												
Gaze-controlled interaction task [65]								•					
Language, visuospatial, and executive processing [66]								•					
Mental arithmetic, short-term memory, aural vigilance [67]								•					

(continued)

Table 2.1 (continued)

Task	BF	BI	BD	FF	FD	SD	SS	PD	PC	PS	IC	ST	HM
Motorbike riding task [68]				•	•	•	•	•					
Ocular following and oral calculation [69]								•		•			
Reading, reasoning, searching, and object manipulation [40]									•				
Simulated/real driving task and mental arithmetic task [70]												•	
Tracking task and mental arithmetic task [71]		•											
Tracking task and mental arithmetic task [72]		•						•				•	
Video game (action-puzzle) task [73]				•	•			•					•
Visual backward masking task [74]								•					
Visual horizontal tracking and visual gauge monitoring [75]	•	•											
Visual search of symbolic displays [39]				•	•			•					
Visuospatial memory task [76]	•		•	•	•	•		•					

Physiological measures *BF* blink frequency, *BI* blink interval/latency, *BD* blink duration, *FF* fixation frequency, *FD* fixation duration, *SD* saccade distance/extent, *SS* saccade speed, *PD* pupil diameter/dilation, *PC* percentage change in pupil size, *PS* power spectrum, *IC* index of cognitive activity, *ST* skin temperature, *HM* head/hand movement

- Digital pen input;
- Gesture input or any other kind of interactive input used to issue system commands.

These responses provide two types of information:

- The inherent content or meaning of the response;
- The manner in which the response was made.

For example, one could type in a sequence of numbers as part of a task in different ways using a variety of equipment – the keys on the top part of the keyboard (above the alphabet), or the keys on the number pad on the right side of the keyboard, or by clicking buttons on a numeric display with a mouse. The string of numbers is the same – this is the content or meaning in the response relevant to the domain task. The manner in which the response is made does not directly affect the outcome of the task, but does provide other information, for example, how long it took to enter the sequence of letters, how much pressure was exerted on each key, and in the case of the mouse usage, features such as the mouse trajectory and the time between clicks [1].

These sources are defined as behavioral rather than performance centric because the information they hold does not directly affect the domain-based outcome of the task, hence there is a lot of margin for differences within and between users. They are objective, and can be collected implicitly, i.e. while the user is completing their task and without overt collection activities (e.g. stopping to ask the user to provide a subjective rating of difficulty), hence suitable for control-room type environments. They are also distinct from physiological measures in that they are mostly or entirely under the user's voluntary control. Some of the measures, for example acoustic speech features, do not fall neatly into the usual definition of "behavioral", however they share with behavioral measures the property of being non-intrusively and continuously acquired, they occur during a task rather than after it, and in most cases they are primarily under partial or full conscious user control, by contrast with post-hoc measures (e.g. performance). While behavioral measures are likely to introduce some variability relative to physiological measures, this variability may turn out to be smaller when the behavior is the response to a task or task type that occurs very often in the user's environment. There is evidence that these kinds of behavioral features can reflect mental states, such as cognitive load. For instance, Gütl et al. [77] used eye tracking to observe subjects' learning activities in real-time by monitoring their eye movements for adaptive learning purposes. Although the visual functions are partly involuntary (e.g. the eye is drawn to salient items in the visual field), gaze is under voluntary control, and can be considered a behavioral measure. Contemporary eye-trackers do not require cumbersome head-sets and can be extracted from video collected through standard webcams. Others have used mouse clicking and keyboard key-pressing behavior to make inferences about participants' emotional states and adapt the system's response accordingly [78, 79].

Previous research also suggests the existence of major speech cues that are related to high cognitive load [12, 80, 81]. Examples of features that have been shown to vary according to task difficulty include pitch, prosody, speech rate, speech energy, and fundamental speech frequency. Some studies have reported an increase in the subjects' rate of speech as well as speech energy, amplitude, and variability under high load conditions [82, 83]. Others have found specific peak intonation [84] and pitch range patterns [83, 85] in high load conditions. Pitch variability has also been shown to potentially correlate to cognitive load [82, 85, 86]. These features are classified as behavioral because they show variations regardless of the meaning of the utterance being conveyed. Additionally, one study employed users' digital-pen gestures and usage patterns to evaluate the usability and complexity of different interfaces [87]. It has been suggested that pen-based interfaces can dramatically improve subjects' ability to express themselves over traditional interfaces because linguistic, alphanumeric and spatial representations bear little cognitive overhead [88].

Higher level features, such as linguistic and grammatical features, may also be extracted from user's spoken language for patterns that may be indicative of high cognitive load. Significant variations in levels of spoken disfluency, articulation rate and filler and pause rates [12] have been found in users experiencing low versus high cognitive load. Extensions of this work attempt to recognize cognitive load levels using a Bayesian network approach [81]; other work has found changes in word frequency and first person plurals [89]. Changes in linguistic and grammatical features have also been used for purposes other than cognitive load measurement [90].

Besides, Verrel et al. [91] showed that cognitive load has significant effect on individual gait patterns during treadmill walking. Therefore, gait patterns also show potential to index cognitive load.

2.6 Estimating Load from Interactive Behavior

Observations of interactive features may be suitable for cognitive load assessment because a user experiencing a high cognitive load will show behavioral symptoms relating to the management of that load. This is likely to suggest a more general effect of an attempt to maximize working memory resources during completion of complex tasks [32, 92]. High cognitive load tasks increase the cognitive demand, forcing more cognitive processes to share fewer resources. It is hypothesized that such reactions will cause changes in interactive and communicative behavior, whether voluntary or otherwise.

The hypothesis that behavioral responses can provide insight into mental states and processing is not without precedent. Spivey et. al. argued that reaching movements made with a computer mouse provide a continuous two-dimensional index of which regions of a scene are influencing or guiding "action plans" – and therefore reflective of changes in cognitive processes [93]. In an experiment involving decision-making, McKinstry et al. found that mouse trajectories for answer selection (YES and NO) options are characterized by the greatest curvature and the lowest peak velocity when the "correct" choice to be made is more ambiguous or more complex [94]. They conclude that spatial extent and temporal dynamics of motor movements can provide insight into high-level cognition [94, 95]. Dale et al. ran a study where participants' hand movements were continuously tracked using a Nintendo® Wii™ remote, as they learned to categorize elements [96]. They noted that participants' arm movements started and finished more quickly and more smoothly (decreased fluctuation and increased perturbation) after learning the categorization rules. The "features of action dynamics" show that participants grow more "confident" over a learning task, and indicate learning has taken place. Van Galen and Huygevoort have shown that time pressure and dual task load results in "biomechanical adaptations of pen pressure" as a coping mechanism

[97]. These studies provide evidence that features of behavioral responses can be harnessed to provide an indication of changes in cognitive processing and coping strategies.

Symptomatic changes in structure, form and quality of communicative and interactive responses are more likely to appear as people are increasingly loaded, as which we will discuss below. With the proliferation of sensor data that can be collected from users through the latest intelligent systems, there is a very specific opportunity to use this behavioural input to detect patterns of change that are correlated with high load, and use these information to guide the adaptation strategies employed by intelligent systems. Here, we use the term "patterns of change" to describe any behavioural change, and that cues are perceptible or observable behaviours that can be used to signal that a change is occurring. Such features have the added advantage of offering an implicit – as opposed to overt – way to collect and assess cues that indicate changes in cognitive load. However, to do this it is necessary to first identify and quantify the fluctuations of features in user interaction as cognitive load varies over a variety of input modalities such as speech and digital-pen input.

The major challenge of choosing the assessment features for automated cognitive load detection is to make sure they satisfy the requirements of consistency, compact representation and automatic acquisition [98]. Cognitive load measurement approaches aim to find effective features which reliably reflect the cues and can be extracted automatically such that they are useful in adaptive systems. They also aim to find a suitable learning or modeling scheme for each index in order to resolve the corresponding level of cognitive load [98]. By manipulating the level of task complexity and cognitive load, and conducting a series of repeated measures user studies in a variety of scenarios, it has been able to identify a series of cognitive load indices based on features from a number of input modalities, specifically, observations of significant changes in speech and digital-pen input that are abstracted from individual application domains in which they occur as well as correlated to high cognitive load. In this book, the term "indices" is used to denote operationalized cues that can be resolved by a machine – and may comprise many individual features (which may or may not be indicative of load on their own).

2.7 Measuring Different Types of Cognitive Load

As discussed in Sect. 2.2, Leppink et al. [18] presented a ten-item instrument to measure three types of cognitive load. Task complexity and the subject's prior knowledge determine the intrinsic load, instructional features that are not beneficial for task completion contribute to extraneous load, and instructional features that are beneficial for task completion contribute to germane load. Ideally, intrinsic load should be optimized in instructional design by selecting tasks that match subject's prior knowledge while extraneous load should be minimized to reduce ineffective load and to allow subject to engage in activities imposing germane load [18]. The data were collected in four lectures and in a randomized experiment in statistics with 0–10 rating scales in [18] for CLM. A ten question subjective survey was used to evaluate three types of cognitive load in the study.

Leppink et al. [99] extended their work in [18] by adapting the survey instrument to the domain of learning languages. Morrison et al. also [100] adapted the Cognitive Load Component Survey in [18] for use in an introductory computer science context. Debue and Leemput [15] exploited eye-tracking to examine the three types of cognitive load by varying the amount of multimedia elements of an online newspaper and examining performance and cognitive load factors. Data analysis results showed the expected opposite relationships between germane and extraneous load, which means that when users perceived the content presentation as unclear, users reported a lower contribution to their learning. Debue and Leemput [15] also showed a positive association between germane load and cognitive absorption (which is defined as "a state of deep involvement with software" [101]) and a non-linear association between intrinsic and germane load.

2.8 Differences in Cognitive Load

2.8.1 Gender Differences in Cognitive Load

Is gender significantly associated with cognitive load levels? For example, do girls and boys show different cognitive load levels during stressful cognitive tasks? This question affects whether approaches to specific human computer interactions need to be varied based on gender. Hwang et al. [102] investigated whether girls have more competition anxiety and exogenous cognitive load than equally able boys during the playing of stressful competitive online games. 235 children in the 6th grade elementary school were recruited for the Chinese radical assembly game (decompose and assemble Chinese characters). It was found that girls did have a higher cognitive load and more competition anxiety from synchronous types of competitive games, but they showed beliefs in technology acceptance constructs that were similar to that of boys. Even with high cognitive load and competition anxiety, the boys and girls didn't show a decrease in their perceived ease of playing and sense of usefulness in using the game to learn Chinese characters for two types of competitive games. Both the boys and girls showed a positive attitude and intentions to play the game [102].

2.8.2 Age Differences in Cognitive Load

Ferreira et al. [103] assessed cognitive load based on psycho-physiological measurements for younger and older adults. Two different types of tasks were used: the

Pursuit Test (PT), in which different curves are entangled and subjects were required to determine where a curve begins and ends, and the Scattered X's test (SX), in which the letter "X" was displayed together with other letters on the screen, subjects were required to locate all the X's in a display. These tests measure perceptual speed and visio-spatial cognitive processing capabilities identified from the fields of psychology and cognitive science. Participants' psycho-physiological responses were recorded with four sensor devices: GSR, ECG (records heart rate and breathing rate (BR)), EEG, an armband (records heat flux – rate of heat transfer on the skin). A total of 128 features were extracted from recorded signals for CLM. It was found that the EEG signal was a better predictor and the BR measurement less important for the older than for the younger participants. Heart rate and breathing rate signals from raw ECG signal were also fairly well represented among the most common features for both age groups, whereas GSR features were seldom used for the younger participants and never for the older participants.

Verrel et al. [91] found that different age groups showed different gait patterns with high cognitive load during treadmill walking. For example, when cognitive load was increased, gait patterns became more regular in those 20–30 years old, less regular in those 70–80 years old, and showed no significant effects in those 60–70 years old.

2.8.3 Static Graphics Versus Animated Graphics in Cognitive Load

In order to examine the role of graphics in learning processes, Dindar et al. [104] developed test questions either with static graphics or with animated graphics accompanied with text to measure students' cognitive load during the learning process. Students' response time, response accuracy, self-reported ratings on cognitive load and secondary task performance were used to measure their cognitive load. It was found that animating graphics increased the response time and secondary task scores (higher secondary task scores refer to higher levels of cognitive load) of the students but did not have any significant effect on their test success. Self-ratings and response accuracy were also found more sensitive to intrinsic cognitive load, whereas response time and secondary task measures were found to be more sensitive to extraneous cognitive load.

2.9 Summary

This chapter presented an overview of the state-of-the-art in cognitive load measurement. Different methods of cognitive load measurement were firstly identified and then the related work of each cognitive load measurement approach was investigated in detail. These include subjective (self-report) measures, physiological measures, performance measures, and behavioral measures. The related work on measuring different types of cognitive load (intrinsic load, extraneous load, and germane load) was also reviewed. It was found that cognitive load may show differences under different conditions, such as gender differences, age differences, and representation differences (static graphics versus animated graphics).

References

1. F. Chen, N. Ruiz, E. Choi, J. Epps, M.A. Khawaja, R. Taib, B. Yin, Y. Wang, Multimodal behavior and interaction as indicators of cognitive load. ACM Trans. Interact. Intell. Syst. **2** (4), 22:1–22:36 (2012)
2. A.D. Baddeley, Working memory. Science **255**, 556–559 (1992)
3. N. Cowan, The magical number 4 in short-term memory: A reconsideration of mental storage capacity. Behav. Brain Sci. **24**(1), 87–114 (2001)
4. J. Sweller, J. Merrienboer, F. Paas, Cognitive architecture and instructional design. Educ. Psychol. Rev. **10**(3), 251–296 (1998)
5. F. Paas, J.E. Tuovinen, H. Tabbers, P.W.M. Van Gerven, Cognitive load measurement as a means to advance cognitive load theory. Educ. Psychol. **38**(1), 63–71 (2003)
6. D. Gopher, R. Braune, On the psychophysics of workload: Why bother with subjective measures? Hum. Factors **26**, 519–532 (1984)
7. R. O'Donnell, F. Eggemeier, Workload assessment methodology, in *Handbook of Perception and Human Performance. Cognitive Processes and Performance, vol. 2*, ed. by K. Boff, L. Kaufman, J. Thomas (Wiley, New York, 1986), pp. 42–1–42–49
8. V.J. Gawron, *Human Performance Measures Handbook* (Lawrence Erlbaum Associates, New Jersey, 2000)
9. F. Paas, P. Ayers, M. Pachman, Assessment of cognitive load in multimedia learning: Theory, methods and applications, in *Recent Innovations in Educational Technology that Facilitate Student Learning*, ed. by D.H. Robinson, G. Schraw (Information Age Publishing Inc, Charlotte, 2008), pp. 11–36
10. P. Chandler, J. Sweller, Cognitive load theory and the format of instruction. Cogn. Instr. **8**(4), 293–332 (1991)
11. D.C. Delis, J.H. Kramer, E. Kaplan, *The Delis-Kaplan Executive Function System* (The Psychological Corporation, San Antonio, 2001)
12. A. Berthold, A. Jameson, Interpreting symptoms of cognitive load in speech input, in *Seventh International Conference on User Modeling (UM99)*, 1999
13. C.S. Ikehara, M.E. Crosby, Assessing cognitive load with physiological sensors, in *Proceedings of the 38th Annual Hawaii International Conference on System Sciences (HICSS 2005)*, Hawaii, USA, 2005
14. F.G.W.C. Paas, J.J.G.V. Merriënboer, Instructional control of cognitive load in the training of complex cognitive tasks. Educ. Psychol. Rev. **6**(4), 351–371 (1994)
15. N. Debue, C. van de Leemput, What does germane load mean? An empirical contribution to the cognitive load theory. Front. Psychol. **5**, 1099 (2014)

16. S. Hart, L. Staveland, Development of the NASA-TLX (Task Load Index): Results of empirical and theoretical research, in *Human Mental Workload*, ed. by P.A. Hancock, N. Meshkati (North Holland Press, Amsterdam, 1988)
17. G. Cierniak, K. Scheiter, P. Gerjets, Explaining the split-attention effect: Is the reduction of extraneous cognitive load accompanied by an increase in germane cognitive load? Comput. Hum. Behav. **25**(2), 315–324 (2009)
18. J. Leppink, F. Paas, C.P.M. Van der Vleuten, T. Van Gog, J.J.G. Van Merriënboer, Development of an instrument for measuring different types of cognitive load. Behav. Res. Methods **45**(4), 1058–1072 (2013)
19. R. Parasuraman, T.B. Sheridan, C.D. Wickens, Situation awareness, mental workload, and trust in automation: Viable, empirically supported cognitive engineering constructs. J. Cogn. Eng. Decis. Making. **2**(2), 140–160 (2008)
20. B. Kerr, Processing demands during mental operations. Mem. Cogn. **1**, 401–412 (1973)
21. N. Marcus, M. Cooper, J. Sweller, Understand instructions. Educ. Psychol. **88**(1), 49–63 (1996)
22. J. Veltman, C. Jansen, *The Role of Operator State Assessment in Adaptive Automation*. TNO-DV3 2005 A245 (TNO Defence, Security and Safety, Soesterbery, 2006)
23. A.J. Byrne, A.J. Sellen, J.G. Jones, Errors on anaesthetic record charts as a measure of anaesthetic performance during simulated critical incidents. Br. J. Anaesth. **80**, 58–62 (1998)
24. G.R.J. Hockey, Operator functional state as a framework for the assessment of performance degradation, in *NATO Advances Research Workshop on Operator Functional State and Impaired Performance in Complex Work Environments*, Il Ciocco, Italy, 2003
25. H.P. Ruffell Smith, *A Simulator Study of the Interaction of Pilot Workload with Errors, Vigilance, and Decisions*. NASA Technical Memorandum (NASA Ames Research Center, Moffett Field, 1979)
26. R. Whelan, Neuroimaging of cognitive load in instructional multimedia. Educ. Res. Rev. **2**(1), 1–12 (2007)
27. J. Moll, R. Zahn, R. de Oliveira-Souza, F. Krueger, J. Grafman, The neural basis of human moral cognition. Nat. Rev. Neurosci. **6**(10), 799–809 (2005)
28. J. Moll, R. De Oliveira-Souza, R. Zahn, The neural basis of moral cognition: Sentiments, concepts, and values. Ann. N. Y. Acad. Sci. **1124**, 161–180 (2008)
29. T. van Gog, F. Paas, N. Marcus, P. Ayres, J. Sweller, The mirror neuron system and observational learning: Implications for the effectiveness of dynamic visualizations. Educ. Psychol. Rev. **21**(1), 21–30 (2008)
30. P. Ayres, F. Paas, Interdisciplinary perspectives inspiring a new generation of cognitive load research. Educ. Psychol. Rev. **21**(1), 1–9 (2009)
31. A.F. Kramer, Physiological metrics of mental workload: A review of recent progress, in *Multiple-Task Performance*, ed. by D.L. Damos (Taylor and Francis, London, 1991), pp. 279–328
32. L.J.M. Mulder, Measurement and analysis methods of heart rate and respiration for use in applied environments. J. Educ. Psychol. **34**(2–3), 205–236 (1992)
33. D. Kennedy, A. Scholey, Glucose administration, heart rate and cognitive performance: Effects of increasing mental effort. Psychopharmacology **149**(1), 63–71 (2000)
34. P. Nickel, F. Nachreiner, Psychometric properties of the 0.1Hz component of HRV as an indicator of mental strain, in *Proceedings of the IEA 2000/HFES 2000: the XIVth Triennial Congress of the International Ergonomics Association and 44th Annual Meeting of the Human Factors and Ergonomics Society*, San Diego, California, 2000
35. R. Brunken, J.L. Plass, D. Leutner, Direct measurement of cognitive load in multimedia learning. Educ. Psychol. **38**(1), 53–61 (2003)
36. G.F. Wilson, C.A. Russell, Real-time assessment of mental workload using psychological measures and artificial neural network. Hum. Factors **45**(4), 635–643 (2003)

37. S.C. Jacobs, R. Friedman, J.D. Parker, G.H. Tofler, A.H. Jimenez, J.E. Muller, P.H. Stone, Use of skin conductance changes during mental stress testing as an index of autonomic arousal in cardiovascular research. Am. Heart. J. **128**(1), 6 (1994)
38. Y. Shi, N. Ruiz, R.Taib, E. Choi, F.Chen.: Galvanic Skin Response (GSR) as an index of cognitive load. In: R. Mary Beth *CHI '07 Extended Abstracts on Human Factors in Computing Systems* 2651–2656
39. R.W. Backs, L.C. Walrath, Eye movement and pupillary response indices of mental workload during visual search of symbolic displays. Appl. Ergon. **23**, 243–254 (1992)
40. S.T. Iqbal, X.S. Zheng, B.P. Bailey, Task-evoked pupillary response to mental workload in human-computer interaction, in *Proceedings of the Internatinal Conference on Computer-Human Interaction (CHI 2004)*, Vienna, Austria, 2004
41. O.V. Lipp, D.L. Neumann, Attentional blink reflex modulation in a continuous performance task is modality specific. Psychophysiology **41**(3), 417–425 (2004)
42. S.P. Marshall, C.W. Pleydell-pearce, B.T. Dickson, Integrating psychological measures of cognitive workload and eye movements to detect strategy shifts, in *Proceedings of 36th Hawaii International Conference on System Sciences (HICSS 2003)*, 2003, vol. 5
43. S. Chen, J. Epps, Automatic classification of eye activity for cognitive load measurement with emotion interference. Comput. Methods. Prog. Biomed **110**(2), 111–124 (2013)
44. W. Wang, Z. Li, Y. Wang, F. Chen, Indexing cognitive workload based on pupillary response under luminance and emotional changes, in *Proceedings of the 2013 International Conference on Intelligent User Interfaces*, 2013, pp. 247–256
45. G. Hossain, M. Yeasin, Understanding effects of cognitive load from pupillary responses using Hilbert analytic phase, in *2014 IEEE Conference on Computer Vision and Pattern Recognition Workshops (CVPRW)*, 2014, pp. 381–386
46. H. Ledger, The effect cognitive load has on eye blinking. Plymouth Stud. Sci. **6**(1), 206–223 (2013)
47. D. Das, D. Chatterjee, A. Sinha, Unsupervised approach for measurement of cognitive load using EEG signals, in *2013 IEEE 13th International Conference on Bioinformatics and Bioengineering (BIBE)*, 2013, pp. 1–6
48. S. Joseph, Measuring cognitive load: A comparison of self-report and physiological methods. PhD thesis, Arizona State University, Arizona, 2013
49. U. Ahlstrom, F.J. Friedman-Berg, Using eye movement activity as a correlate of cognitive workload. Int. J. Ind. Ergon. **36**(7), 623–636 (2006)
50. J.B. Brookings, G.F. Wilson, C.R. Swain, Psychophysiological responses to changes in workload during simulated air traffic control. Biol. Psychol. **42**(3), 361–377 (1996)
51. L.L. Di Stasi, M. Marchitto, A. Antolí, T. Baccino, J.J. Cañas, Approximation of on-line mental workload index in ATC simulated multitasks. J. Air Transp. Manag. **16**(6), 330–333 (2010)
52. M. Guhe, W.D. Gray, M.J. Schoelles, W. Liao, Z. Zhu, Q. Ji, Non-intrusive measurement of workload in real-time. Proc. Hum. Factors. Ergonomics. Soc. Annu. Meet. **49**(12), 1157–1161 (2005)
53. N.L. Thomas, Y. Du, T. Artavatkun, J. She, Non-intrusive personalized mental workload evaluation for exercise intensity measure, in *Digital Human Modeling*, ed. by V.G. Duffy (Springer, Berlin/Heidelberg, 2009), pp. 315–322
54. M.A. Recarte, E. Pérez, A. Conchillo, L.M. Nunes, Mental workload and visual impairment: Differences between pupil, blink, and subjective rating. Span. J. Psychol. **11**(2), 374–385 (2008)
55. T. de Greef, H. Lafeber, H. van Oostendorp, J. Lindenberg, Eye movement as indicators of mental workload to trigger adaptive automation, in *Foundations of Augmented Cognition. Neuroergonomics and Operational Neuroscience*, ed. by D.D. Schmorrow, I.V. Estabrooke, M. Grootjen (Springer, Berlin/Heidelberg, 2009), pp. 219–228
56. H. J. Veltman, W.K. Vos, Facial temperature as a measure of operator state, in *Proceedings of the 11th International conference on Human-computer interaction*, 2005

57. A. Murata, H. Iwase, Evaluation of mental workload by fluctuation analysis of pupil area, in *Proceedings of the 20th Annual International Conference of the IEEE Engineering in Medicine and Biology Society, 1998*, 1998, vol. 6, pp. 3094–3097
58. Y.-F. Tsai, E. Viirre, C. Strychacz, B. Chase, T.-P. Jung, Task performance and eye activity: Predicting behavior relating to cognitive workload. Aviat. Space Environ. Med. **78**(5 Suppl), B176–B185 (2007)
59. M. Schwalm, A. Keinath, H.D. Zimmer, Pupillometry as a method for measuring mental workload within a simulated driving task, in *Human factors for assistance and automation*, ed. by D. de Waard, N. Gérard, L. Onnasch, R. Wiczorek, D. Manzey (Shaker Publishing, Maastricht, 2008), pp. 1–13
60. O. Palinko, A.L. Kun, A. Shyrokov, P. Heeman, Estimating cognitive load using remote eye tracking in a driving simulator, in *Proceedings of the 2010 Symposium on Eye-Tracking Research & Applications*, New York, NY, USA, 2010, pp. 141–144
61. Y. Zhang, Y. Owechko, J. Zhang, Driver cognitive workload estimation: a data-driven perspective, in *The 7th International IEEE Conference on Intelligent Transportation Systems, 2004. Proceedings*, 2004, pp. 642–647
62. B.P. Bailey, S.T. Iqbal, Understanding changes in mental workload during execution of goal-directed tasks and its application for interruption management. ACM. Trans. Comput. Hum. Interact. **14**(4), 21:1–21:28 (2008)
63. J.A. Veltman, A.W. Gaillard, Physiological workload reactions to increasing levels of task difficulty. Ergonomics **41**(5), 656–669 (1998)
64. G.F. Wilson, An analysis of mental workload in pilots during flight using multiple psychophysiological measures. Int. J. Aviat. Psychol. **12**(1), 3–18 (2002)
65. M. Pomplun, S. Sunkara, Pupil dilation as an indicator of cognitive workload in human-computer interaction, in *Proceedings of the International Conference on HCI 2003*, 2003
66. M.A. Just, P.A. Carpenter, A. Miyake, Neuroindices of cognitive workload: Neuroimaging, pupillometric and event-related potential studies of brain work. Theor. Issues. Ergonomics Sci. **4**, 56–88 (2003)
67. J. Klingner, R. Kumar, P. Hanrahan, Measuring the task-evoked pupillary response with a remote eye tracker, in *Proceedings of the 2008 Symposium on Eye Tracking Research & Applications*, New York, NY, USA, 2008, pp. 69–72
68. L.L. Di Stasi, V. Álvarez-Valbuena, J.J. Cañas, A. Maldonado, A. Catena, A. Antolí, A. Candido, Risk behaviour and mental workload: Multimodal assessment techniques applied to motorbike riding simulation. Transport. Res. F: Traffic Psychol. Behav. **12**(5), 361–370 (2009)
69. M. Nakayama, Y. Shimizu, Frequency analysis of task evoked pupillary response and eye-movement, in *Proceedings of the 2004 Symposium on Eye Tracking Research & Applications*, New York, NY, USA, 2004, pp. 71–76
70. C.K. Or, V.G. Duffy, Development of a facial skin temperature-based methodology for non-intrusive mental workload measurement. Occup. Ergon. **7**(2), 83 (2007)
71. K. Ryu, R. Myung, Evaluation of mental workload with a combined measure based on physiological indices during a dual task of tracking and mental arithmetic. Int. J. Ind. Ergon. **35**(11), 991–1009 (2005)
72. L. Wang, V.G. Duffy, Y. Du, A composite measure for the evaluation of mental workload, in *Digital Human Modeling*, ed. by V.G. Duffy (Springer, Berlin/Heidelberg, 2007), pp. 460–466
73. T. Lin, A. Imamiya, Evaluating usability based on multimodal information: An empirical study, in *Proceedings of the 8th International Conference on Multimodal Interfaces*, New York, NY, USA, 2006, pp. 364–371
74. S.P. Verney, E. Granholm, D.P. Dionisio, Pupillary responses and processing resources on the visual backward masking task. Psychophysiology **38**(1), 76–83 (2001)
75. D.L. Neumann, Effect of varying levels of mental workload on startle eyeblink modulation. Ergonomics **45**(8), 583–602 (2002)

76. K.F. Van Orden, W. Limbert, S. Makeig, T.P. Jung, Eye activity correlates of workload during a visuospatial memory task. Hum. Factors **43**(1), 111–121 (2001)
77. C. Guetl, M. Pivec, C. Trummer, V. M. Garcabarrios, F. Mdritscher, J. Pripfl, M. Umgeher, ADELE (adaptive e-learning with eye-tracking): Theoretical background, system architecture and application scenarios. Eur. J. Open, Distance. E-Learning, vol. 2, 1–16 2005
78. W.S. Ark, D.C. Dryer, D.J. Lu, The emotion mouse, in *Proceedings of HCI International (the 8th International Conference on Human-Computer Interaction) on Human-Computer Interaction: Ergonomics and User Interfaces*, ed. by Z. Bullinger, 1st edn. (Lawrence Erlbaum Association, London, 1999), pp. 818–823
79. J. Liu, E. Al, An adaptive user interface based on personalised learning. IEEE Intell. Syst. **18** (2), 52–57 (2003)
80. H. Keraenen, E. Vaeyrynen, R. Paeaekkoenen, T. Leino, P. Kuronen, J. Toivanen, T. Seppaenen, Prosodic features of speech produced by military pilots during demanding tasks, in *Proceedings of the Fonetiikan Paeivaet 2004*, 2004
81. C. Mueller, B. Grossmann-hutter, A. Jameson, R. Rummer, F. Wittig, Recognising time pressure and cognitive load on the basis of speech: An experimental study, in *Proceedings of the Eighth International Conference on User Modeling (UM 2001)*, Berlin, 2001
82. M. Brenner, T. Shipp, E. Doherty, P. Morrissey, Voice measures of psychological stress: Laboratory and field data, in *Vocal Fold Physiology, Biomechanics, Acoustics, and Phonatory Control*, ed. by I. Titze, R. Scherer (Denver Center for the Performing Arts, Denver/Colorado, 1985), pp. 239–248
83. E. Lively, D.B. Pisoni, W.V. Summers, R. Bernacki, Effects of cognitive workload on speech production: Acoustic analyses and perceptual consequences. J. Acoust. Soc. Am. **93**, 2962–2973 (1993)
84. S. Kettebekov, Exploiting prosodic structuring of coverbal gesticulation, in *ICMI'04: 6th international conference on Multimodal interfaces*, State College, PA, USA, 2004, pp. 105–112
85. C. Wood, K. Torkkola, S. Kundalkar, Using driver's speech to detect cognitive workload, in *Proceedings of the 9th Conference on Speech and Computer (SPECOM 2004)*, France, 2004
86. E.J. Tolkmitt, K.R. Scherer, Effect of experimentally induced stress on vocal parameters. J. Exp. Psychol. **12**, 302–312 (1986)
87. S. Oviatt, Human-centered design meets cognitive load theory: Designing interfaces that help people think, in *Proceedings of the 14th annual ACM international conference on Multimedia (MULTIMEDIA 2006)*, 2006
88. S. Oviatt, Designing interfaces that stimulate ideational superfluency, in *Proceedings of the INKE 2009: Research Foundations for Understanding Book and Reading in the Digital Age: Implementing New Knowledge Environments*, 2009
89. J.B. Sexton, R.L. Helmreich, Analyzing cockpit communication: The links between language, performance, error, and workload. J. Hum. Perform. Extrem. Environ. **5**(1), 63–68 (2000)
90. J. Schilperoord, On the cognitive status of pauses discourse production, in *Contemporary Tools and Techniques for Studying Writing*, ed. by T. Olive, C.M. Levy (Kluwer Academic Publishers, London, 2001)
91. J. Verrel, M. Lövdén, M. Schellenbach, S. Schaefer, U. Lindenberger, Interacting effects of cognitive load and adult age on the regularity of whole-body motion during treadmill walking. Psychol. Aging **24**(1), 75–81 (2009)
92. S. Oviatt, R. Coulston, .R. Lunsford, When do we interact multimodally?: Cognitive load and multimodal communication patterns, in *Proceedings of the 6th international conference on Multimodal interfaces (ICMI 2004)*, New York, USA, 2004
93. M.J. Spivey, M. Grosjean, G. Knoblich, Continuous attraction toward phonological competitors. Proc. Natl. Acad. Sci. U S A **102**(29), 10393–10398 (2005)
94. C. Mckinstry, R. Dale, M.J. Spivey, Action dynamics reveal parallel competition in decision making. Psychol. Sci. **19**(1), 22–24 (2008)

95. D.A. Rosenbaum, The Cinderella of psychology: The neglect of motor control in the science of mental life and behavior. Am. Psychol. **60**, 308–317 (2005)
96. R. Dale, J. Roche, K. Snyder, R. Mccall, Exploring action dynamics as an index of paired-associate learning. PLoS ONE **3**(3) (2008). e1728
97. G.P. Galen, M. Van Huygevoort, Error, stress and the role of neuron-motor noise in space oriented behavior. Biol. Psychol. **51**, 151–171 (2000)
98. B. Yin, F. Chen, N. Ruiz, E. Ambikairajah, Speech-based cognitive load monitoring system, in *IEEE International Conference on Acoustics, Speech and Signal Processing, 2008. ICASSP 2008*, 2008, pp. 2041–2044
99. J. Leppink, F. Paas, T. van Gog, C.P.M. van der Vleuten, J.J.G. van Merriënboer, Effects of pairs of problems and examples on task performance and different types of cognitive load. Learn. Instr. **30**, 32–42 (2014)
100. B.B. Morrison, B. Dorn, M. Guzdial, Measuring cognitive load in introductory CS: Adaptation of an instrument, in *Proceedings of the Tenth Annual Conference on International Computing Education Research*, 2014, pp. 131–138
101. R. Agarwal, E. Karahanna, Time flies when you're having fun: Cognitive absorption and beliefs about information technology usage1. MIS Q. **24**(4), 665–694 (2000)
102. M.-Y. Hwang, J.-C. Hong, H.-Y. Cheng, Y.-C. Peng, N.-C. Wu, Gender differences in cognitive load and competition anxiety affect 6th grade students' attitude toward playing and intention to play at a sequential or synchronous game. Comput. Educ. **60**(1), 254–263 (2013)
103. E. Ferreira, D. Ferreira, S. Kim, P. Siirtola, J. Roning, J.F. Forlizzi, A. K. Dey, Assessing real-time cognitive load based on psycho-physiological measures for younger and older adults, in *2014 IEEE Symposium on Computational Intelligence, Cognitive Algorithms, Mind, and Brain (CCMB)*, 2014, pp. 39–48
104. M. Dindar, I. Kabakçı Yurdakul, Measuring cognitive load in test items: Static graphics versus animated graphics. J. Comput. Assist. Learn. **31**(2), 148–161 (2014)

Chapter 3
Theoretical Aspects of Multimodal Cognitive Load Measures

The overall theme of this chapter is to highlight important aspects of the theory underlying the development of Multimodal Cognitive Load Measures (MCLM). Here we bring forth elements that have historically played an important role in creating the need and the framework for robust multimodal cognitive load measures. We begin by discussing briefly how cognitive load investigations became critical to the area of human performance studies. Then we discuss cognitive load as a psychological construct and how it evolved within the psychology of learning under the umbrella of Cognitive Load Theory (CLT). Here the modality principle of CLT's sister theory, namely, the Cognitive Theory of Multimedia Learning (CTML) deserves a special mention and has been discussed. Following this, we look at mathematical issues related to analysing working memory capacity – especially the statistical measures for workload capacity analysis. There follows a section detailing issues in experimental testing and the evaluation environment, with focus on dual-task methodologies of inducing load on working memory. Throughout these sections, we focus on how the theoretical underpinnings of various approaches were forced to evolve in favour of multimodality, as repeated attempts were made to measure cognitive load in an increasingly direct and precise manner. Finally we make our case for the data driven multimodal cognitive load approach and discuss the implications for human decision-making and trust research. We further hope that this multimodal framework can significantly further the cause of Bayesian models ([1–3]) of cognition as well as more recently Quantum models [4] of cognition by providing them with a framework for empirical/sensory external link to user interaction modalities.

F. Chen et al., *Robust Multimodal Cognitive Load Measurement*, Human–Computer Interaction Series, DOI 10.1007/978-3-319-31700-7_3

3.1 Load? What Load? Mental? Or Cognitive? Why Not Effort?

Terms like cognitive load, mental effort, cognitive workload and more recently cognitive effort continue to be used by researchers – extensively and variedly – in several overlapping research domains. They are psychological constructs that help make sense of human performance, learning and other cognitive processes. Here we try to establish some common themes that will help us through the ensuing discussion.

Broadly speaking, cognitive load (in the psychology of learning) has traditionally meant the load experienced by working memory when humans engage in a variety of cognitively intensive tasks ranging from learning to decision-making and more. However, this idea of cognitive load was superseded by a popular psychological construct 'mental load', defined by Moray [5] in the human factors psychological domain. The idea of mental load was based on the difference between the task demands and the person's ability to master those demands. The focus of these studies was to identify situations that may result in critical operator overload. Some researchers have worked with mental 'effort' as opposed to the more general term mental load. This type of usage is thought to get better subjective evaluations from users, but is not broad enough to include factors like environment and task complexity whenever a more comprehensive scenario needs to be evaluated. Then again, more recently, we see a trend amongst neuro-economic inclined researchers to label the 'intensity of engagement' in cognitive related activities as 'cognitive effort' [6]. One possible implication arising from this usage is the possibility of linking intensity of engagement to predictable intuitive/rational type of decision making. We discuss this later in Sect. 3.5 of this chapter.

Since most models of working memory accept some form of limitation on its capacity and resources – the goal of many cognitive/mental load studies is to 'optimally manage' the load on working memory generally over either short or extended periods of time since both overload and underload can be considered undesirable. Overload (sometimes also referred to as stress) can lead to substandard performance and even fatal errors in mission-critical environments, while underload can result in similar sub-standard performance, but due to lack of motivation or boredom. Managing cognitive load (on working memory) is not just critical for efficient learning and effective decision-making but for the proper functioning of almost all cognitive processes.

3.2 Mental Load in Human Performance

In this section we look at some of the foundational studies of mental load in human (or *operator*) performance, some research efforts with respect to mental workload scales and curves and finally a few words about the 'red lines' to be found within

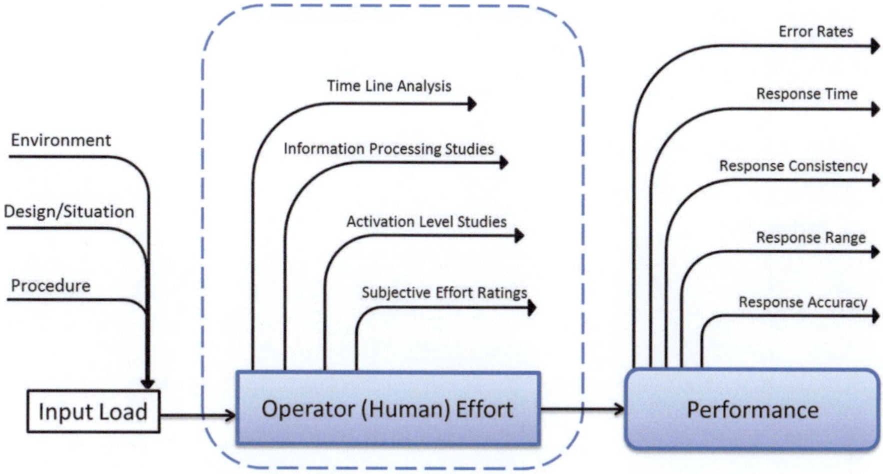

Fig. 3.1 Attributes of operator workload (Adapted from [5])

cognitive (and physical) workloads. One key theme running throughout this section is the significant realization that there can be no single measure that can be recommended as the definitive measure of mental load. We learn from all these experiences and find extensive support for our use of multimodality to measure cognitive load.

A key point to note in this section is the explicit use of word 'operator', often used interchangeably with 'human'. The reason for this is the more recent twenty first century focus on 'human' performance resulting from the explosion of user-friendly interfaces and digital devices meant for all humans and not just specialists like 'operators' (a term that carried historical baggage of the industrial era's heavy machinery engineering operations). In the spirit of giving due credit to previous research, we use the term 'operator' performance when discussing older research, but then switch to 'human' performance when the context becomes more recent.

3.2.1 Mental Workload: The Early Years

Studies of mental load in human performance have a long and rich history. Here we would like to mention some contributions of Neville Moray, Robert Williges and Walter Wierwille. Moray's [5] edited collection, 'Mental Workload: Its Theory and Measurement' is a classic that identifies and predicts almost all issues relevant in the study and measurement of mental workload today. In chapter one of this book, Gunnar Johannsen declares the attributes of operator workload to be broadly divided into (a) input load, (b) operator effort, and (c) performance. Details can be seen in Fig. 3.1. Johannsen then [7] argued operator effort to be a function of

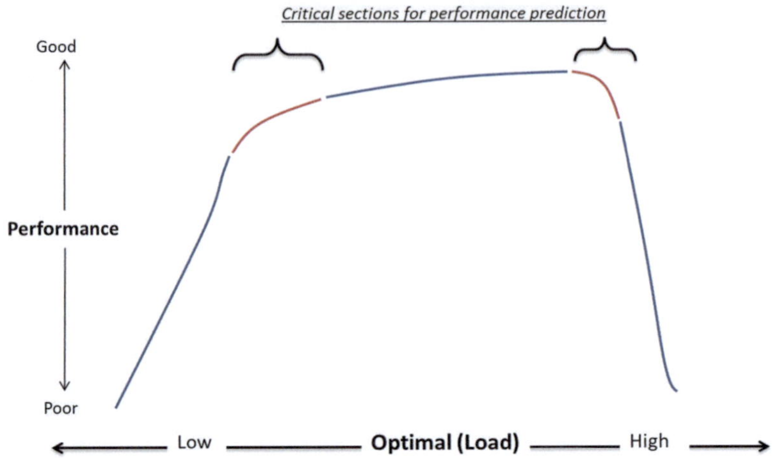

Fig. 3.2 Relation between operator performance and input load (Adapted from [7])

input load, operator state and internal performance criteria, and capable of being modelled by the expression:

$$\text{Effort} = f(\text{Load, Operator-State, Internal Performance Criteria}).$$

However, since the operator state could depend on a number of factors including general background, personality, motivation, attentiveness etc., it remains a challenge to generalize over population variance. Furthermore the terms of stress and strain were used interchangeably with load and effort respectively. Sometimes the term stress was also used as a convenient replacement for peak load. Back then, workload was an umbrella concept which included input load and operator effort. Measures of input load included environmental, procedural, and design or situational variables. Whereas, measures of operator effort are classified into four groups: (1) time-line analysis (where the execution times of task elements are assessed as well as total time needed), (2) information processing studies (where secondary tasks are used to measure spare mental capacity using the assumption of limited channel capacity), (3) operator activation-level studies (based on the hypothesis that the level of physiological activity depends on an operator's effort), and (4) subjective effort ratings (self-assessments by operators but limited to overall assessment only).

A critical point to note is that the four measurement techniques mentioned above differ in the sense that they measure different aspects of effort. For example, (a) time-line analysis is load oriented, (b) information-processing studies are performance oriented, (c) operator activation-level studies are operator-state-oriented, and (d) subjective effort ratings assess only conscious aspects of effort. This clearly implies that only a small portion of the multi-faceted area of human operator workload can be assessed by any single measure.

Johannsen [7] also presented a hypothetical relationship between operator performance and input load on multi-task procedures (see Fig. 3.2). As long as time required (T_R) matches the time available (T_A), a reasonable good performance level can be predicted for 'optimal' load. The two curve sections labelled 'areas of need for prediction capability' are very interesting for real-time monitoring of cognitive load features and have been made use of in real-time approach to detecting changes in cognitive load (more details in Chap. 15 in this book).

Interestingly, in the same year of 1979, two reviews ([8, 9]) were published summarizing the state of the art in physiological and behavioural measures of aircrew mental load. Physiological measures reviewed included Flicker Fusion Frequency (FFF), Critical Fusion Frequency (CFF), Galvanic Skin Response (GSR), skin impedance, Electrocardiogram (ECD), phonocardiogram, plethysmogram, heart rate, heart rate variability, blood pressure, Electromyogram (EMG), muscle tension, Electroencephalogram (EEG), Evoked Cortical Potentials (ECP), eye and eyelid movement, pupillary dilation, respiration analysis, body fluid analysis and speech pattern analysis. The underlying concept in physiological monitoring was that involuntary changes also take place in the physiological processes of the human body (eg body chemistry, nervous system activity, circulatory or respiratory activity) as operator workload changes. Thus workload maybe assessed by measuring and processing the appropriate physiological variables. Another similar idea was that high workload levels are accompanied by increased emotional stress. And stress can then be measured by physiological recordings and thus acts like an intermediate variable. However, it was also understood as a limitation that operator behaviour (eg physical exertion) other than mental workload could also have an effect on physiological measures. Finally the review explicitly concluded that no single physiological measure appeared to be a strong indicator in itself, but must be combined with other workload assessment techniques to provide a more complete understanding of workload.

In the second review of [9], 14 behavioural measures were assessed and grouped into three broad categories of subjective opinions, spare mental capacity and ones that were primarily task oriented. *Subjective opinions* had the disadvantage of being very general and only carried out after the task. From the *spare mental capacity* group, only task component/time summation and time estimation techniques appeared to be promising. However, the major limitation here was the basic assumption of constant workload capacity in operators – though there exists evidence that human workload capacity can change in times of stress or intense motivation. We will discuss this point further in sections ahead. Finally that leaves methods using 'measures of spare mental capacity' and these can be expected to be in error by roughly the same amount as the fluctuation of the limit of working memory.

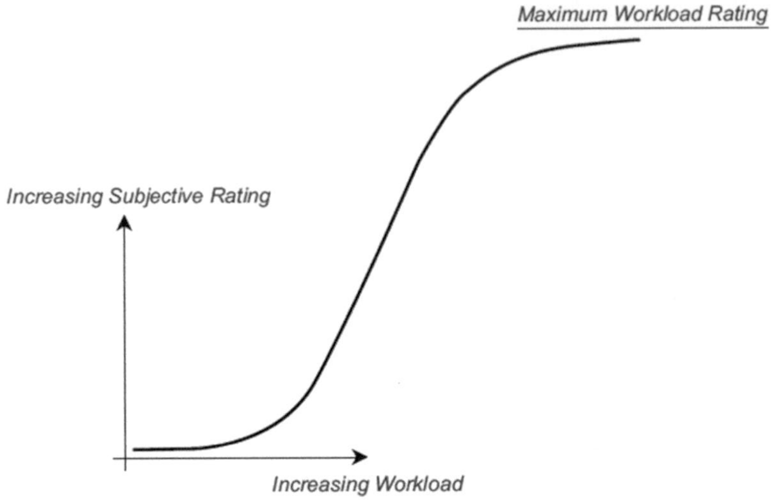

Fig. 3.3 Subjective workload curve ([14], copyright © 2015 SAGE, reprinted by permission of SAGE Publications, Inc.)

3.2.2 Subjective Mental Workload Scales and Curve

Moving on, Moray became convinced that subjective mental workload is of increasing importance in user-machine systems, as the role of the human operator becomes less to control and more to monitor complex systems. He argued extensively about the nature of time stress trying to understand the origins of subjective feelings of load. Several subjective scales were developed during this period. Here we briefly mention two of the most commonly used techniques of subjective mental workload assessment.

First came the NASA-Task Load Index (NASA-TLX; [10]), which included six subscales exploring the Mental Demand, Physical Demand, Temporal Demand, Own Performance, Effort, and Frustration Level. Second, the Subjective Workload Assessment Technique (SWAT; [11]) describes three dimensions of operator workload: Time Load, Mental Effort Load and Psychological Stress Load. These two subjective measurement techniques have been suggested to be relatively similar [12]; especially where Time Load corresponds to Temporal Demand dimensions; Mental Effort Load to Mental Demand and Effort dimensions; and the Psychological Stress Load to Frustration dimensions. Both techniques are largely used in the field of aeronautics, as shown for instance in a study by Collet et al. [13] that revealed a positive correlation between the number of aircraft to control and the NASA-TLX score provided by air traffic controllers. Furthermore, controllers self-rated workload closely paralleled the change in the number of aircraft to be controlled, indicating a high sensitivity of NASA-TLX to small workload changes.

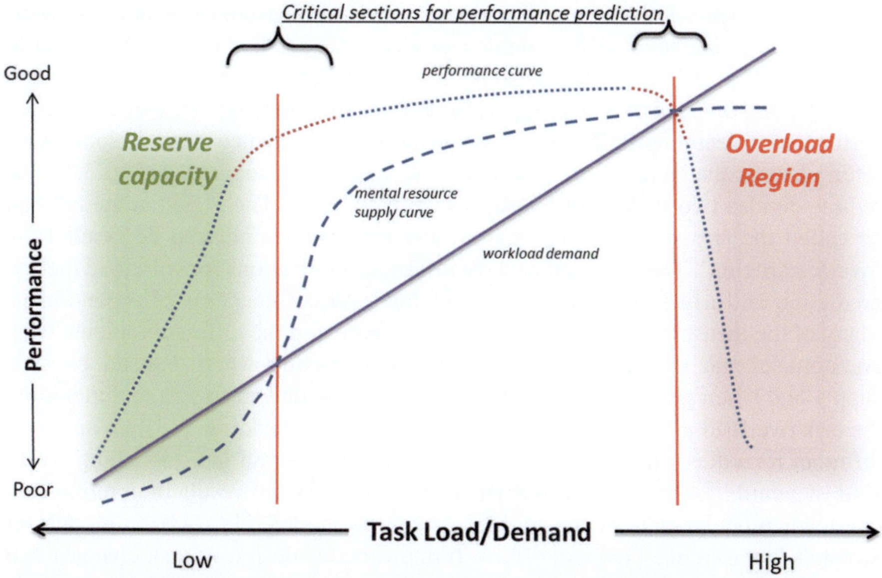

Fig. 3.4 Performance prediction dynamics and red lines (Adapted/inspired from [7], [15])

Results from subjective mental workload assessment are often interpreted linearly and incorrectly. Estes [14] argued that ratings of subjective mental workload increase nonlinearly with a unitary increase in memory working load (see Fig. 3.3). He supported this idea with experimental evidence. We discuss this point later in the chapter, but for the moment it will suffice to note that this indicates a general tendency in users to maintain a reasonable status quo of performance in the face of minor variations in workload demand. The user may be consciously ignoring (or unconsciously tolerating) these minor variations in workload demand until a personal threshold is reached. Beyond this threshold, the user subjectively assesses personal effort to have changed levels.

3.2.3 *Cognitive Workload and Physical Workload Redlines*

The challenge of assessing and measuring mental workload derives from a fundamental theoretical concern for an understanding of cognition, then further computing the cost of neurocognitive understanding of actions within the brain. Thus, mental workload assessment lies very much within both the cognitive revolution of psychology but is also encompassed by the more recent developments in neuroergonomics. However, alongside these theoretic scientific concerns, the need to assess mental work is also fueled by the practical necessity to measure mental activity and allocate tasks in the modern electronic workplace.

In any resource-limited system, the most relevant measure of demand is specified relative to the supply of available resources. We see this relationship conceptualized in Fig. 3.4, but if we consider the vertical axis as representing resource supply, when demand exceeds supply, further demand increases will lead to further performance decrements. The break point on the performance curve is sometimes referred to as the 'redline' of workload [16, 17], and is marked in Fig. 3.4. The redline divides two regions of the supply demand space. The region at the left can be called the 'reserve capacity' region, and that to the right can be labeled the 'overload region'. The two regions have different implications for workload theory, prediction and assessment, as well as for the kinds of concerns of ergonomists. Many of the measures are also differentially sensitive in the different regions. Both ergonomists and designers are interested in predicting when demand exceeds supply and when performance declines as a result, in understanding and modeling the task overload management strategies used (eg task shedding; [18]), in applying different remedies when this overload condition occurs and in establishing workload standards. When this performance decrement is the result of multitasking overload, models such as the multiple resource model ([19–21]) or models of crosstalk interference [18] can offer a framework for design or task changes that will reduce the demand and resulting degradation of performance. This may include using separate, rather than common resources; it may include reducing the resource demands of the task (eg by reducing working memory load, or automating parts of the task), extensive training to expertise, reassigning some of the tasks to another operator or changing procedures in such a way that previously concurrent tasks can now be performed sequentially. These latter solutions also derive from any resource model (single or multiple). More details about the current state of mental workload science in ergonomics can be found in [15].

3.3 Cognitive Load in Human Learning

Detailed expositions of CLT and a CTML are presented by Plass et al. [22], Sweller et al. [23] and Mayer [24–26]. Here we selectively summarize the evolutionary theoretical background for CLT and then go on to discuss how attempts to measure cognitive load directly have helped these theories evolve. Also of interest for us is the treatment of modalities in CTML.

John Sweller can rightly be labelled as the 'father of cognitive load theory' as he and his colleagues have worked diligently over the last decades to develop CLT into a viable instrument of instructional design. Sweller et al. [23] makes a detailed case for CLT by embedding it within biological evolutionary theory. In this framework, information is classified into two categories. The first category is called biologically primary knowledge and consists of knowledge that humans have specifically evolved to acquire naturally (eg learning of first language by a child). Whereas, the second category is biologically secondary knowledge that humans need for cultural reasons but have not specifically evolved to acquire naturally. Educational institutions exist to cater to this need and it is here that CLT has exclusive application. In other words, Sweller et al. [23] have provided an evolutionary account of a natural information processing system that can be extended to explain how information is processed for human cognition. Although CLT uses an evolutionary framework to emphasize the existing human cognitive architecture, the evolutionary nature of these capacities is not its main focus. The ultimate aim of CLT is to use the existing knowledge of human cognitive processes to provide effective instructional design principles. When processing biologically secondary information, human memory includes a working memory that is limited in capacity and duration if dealing with novel information, but effectively unlimited in capacity and duration if dealing with familiar information previously stored in the very large long-term memory. Therefore learning instructions need to consider the limitations of working memory so that information can be successfully transferred into long-term memory. Once appropriate information is stored in the long-term memory, the capacity and duration limits of working memory no longer apply, and the tasks that were initially (perhaps) impossible or inconceivable can become part of normal routine.

The goals of learning are achieved with the construction and automation of schemas. According to CLT, multiple elements of information can be chunked as single elements when fitted into cognitive schemas, which can then be, to a large extent, automated. These automated schemas can bypass working memory during mental processing thereby circumventing associated bottlenecks of this type of memory. So, in summary, the aim of CLT (as a theory of instructional design) is 'to facilitate the acquisition of knowledge in long term memory via a working memory that is limited in capacity and duration until it is transformed by knowledge held in long term memory' (Sweller et al. [23], p. vii).

Cognitive load imposed on working memory by various instructional procedures originates from either the intrinsic nature of the material (*intrinsic load*) to be processed or from the manner in which material is presented and the activities required of the learners (*extrinsic load*). The key for CLT as a learning theory is to design instructional procedures that minimize extrinsic load and manage intrinsic load in such a manner that leads to optimal learning (which for CLT researchers, reflects the productive or *germane load*). With this in mind, we find that to design instructional procedures aligned to the limitations of human working memory; there needs to be ways of measuring cognitive load directly and then trying to manage it; or investigating the adverse effects resulting from complex instructional designs and then eliminating them. Traditional CLT has largely remained focussed on the later approach and gathered a large body of cognitive load effects (eg goal-free effect, worked example & problem completion effect, split-attention effect, modality effect etc.). Next we look at the stages of CLT and its attempts to measure cognitive load.

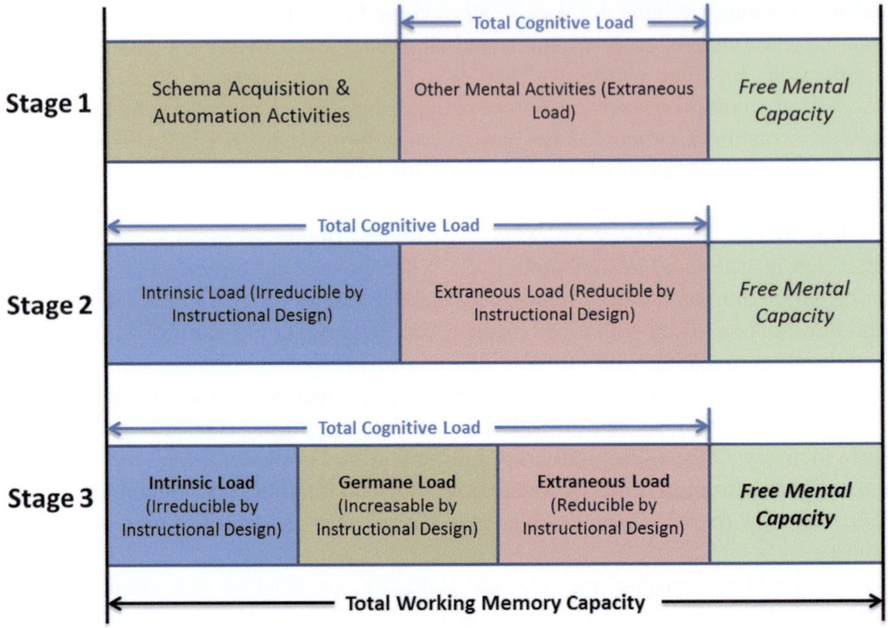

Fig. 3.5 Stages of cognitive load theory (Adapted from [27])

3.3.1 Three Stages of CLT: The Additivity Hypothesis

Traditional CLT focussed on the relation between the type of cognitive processes elicited by different problem-solving methods and schema acquisition. Schema acquisition is the building block of skilled performance and requires attention directed to problem states and their associated solutions. Learning is enhanced with schema acquisition (the details of schema acquisition can be found in Sect. 3.3.2). Other cognitive activities must remain limited to avoid putting heavy cognitive load that would interfere with learning. In the early stages (ie Stage 1) of CLT research, the primary focus was 'other mental activities' (see Fig. 3.5) referred to as extraneous load and that was thought to be reducible by better instructional design. With stage two, there was a realization that the inherent difficulty of the task could not be reduced by instructional design. This gave rise to the concept of intrinsic load that corresponded to element interactivity within task and was considered the additive constituent (along with extrinsic load) for total cognitive load. Finally, in the third stage, another additive component, the concept of germane load, was introduced as the positive/productive load that could be increased depending on the capacity and capability of the learner.

The additive nature of these three types of load remains a problematic assumption for many researchers. It is theoretically convenient to attribute the three load types to different sources but empirically quite challenging to measure all three separately. There could be other non-exclusive ways of interaction between the load types but these have yet to be studied.

3.3.2 Schema Acquisition and First-in Method

Schemas are the units of knowledge representation that allow us to treat elements of information as larger, higher level chunks, thereby reducing capacity demands on working memory and allowing efficient use of the basic information processing features of our cognitive architecture. Cognitive mechanisms of schema acquisition and their transfer from consciously controlled to automatic processing are the major learning mechanisms and foundations of our intellectual abilities and skilled performance. As per Kalyuga [22], schemas represent knowledge as stable patterns of relationships between elements. They describe classes of structures that are abstracted from specific instances and are used to categorize such instances. Multiple schemas can be linked together and organized into larger hierarchical structures. Such organized knowledge structures are a major mechanism for extracting meaning from information, acquiring and storing knowledge in long-term memory, bypassing the limitations of working memory, increasing the strength of memory, guiding the retrieval and recall of information and providing connections to prior knowledge.

Schematic knowledge structures can be empirically evaluated by using grouping and categorizing tasks, by using problems with ambiguous material in their statements or using the 'text' editing technique. Cognitive science methods used for diagnosing individual knowledge structures are based on interviews, think-aloud procedures and different forms of retrospective report. However, more recently rapid online methods of cognitive diagnosis of organized knowledge structures are based on the assumption that if schemas in long-term memory alter the characteristics of working memory, than tracing the immediate content of working memory during task performance may provide a measure of levels of acquisition of the corresponding schematic knowledge structure. The idea is known as the 'first-step' method. In this, learners are presented with a task for a limited time and then asked to rapidly indicate their first step towards the solution of the task. For learners with different levels of expertise their first step could involve different cognitive activities – an expert may immediately provide the final answer whereas the novice may start attempting some random solution search.

Load detection strategies based on the first-step method can help predict changed or sub-optimal behaviour just before a full-fledged decline in performance, or at least place the system on high alert for an imminent probable decline.

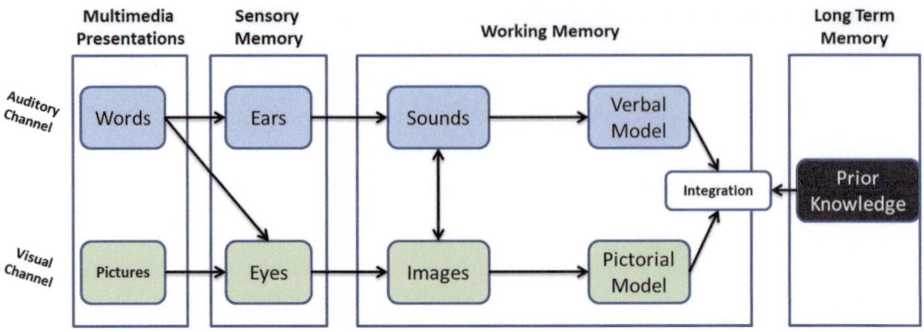

Fig. 3.6 Cognitive theory of multimedia learning (Adapted from [28])

3.3.3 *Modality Principle in CTML*

Multimedia allows the combination of different presentation formats, such as pictures, animations, text, or music in flexible ways via different sensory modalities. Many media designers assume that multimedia always allows for better learning and understanding however this has been shown to be far from truth. Multimedia is also expected to motivate learners, thus increasing their invested cognitive effort, leading to better learning.

Mayer [26] assumed that two sensory subsystems exist in working memory – an auditory systems and a visual system. Furthermore, these two channels have limited capacities to convey and process information. And thirdly, humans are active sense makers – they engage in active cognitive processing to construct coherent knowledge structures from available external information and their prior knowledge. Active learning includes five coordinated activities: (1) selecting relevant words, (2) selecting relevant images, (3) organizing selected words into a verbal mental model, (4) organizing selected images into a pictorial mental model, and (5) integrating the verbal and pictorial model with prior knowledge into a coherent mental representation (Fig. 3.6).

However, multimedia principles do not apply under all conditions. If a learner's prior knowledge is high, learning from a single medium can lead to better learning results than multimedia learning [29]. Multimedia principles also do not apply if the learner possesses insufficient spatial abilities to process the information. There are several guidelines provided concerning specific instructional design characteristics of multimedia learning. Of particular interest to us are modality studies. The modality principle postulates that students learn better from animation and narrated text that from animation and written text. Using two channels expands the effective working memory capacity and reduces the probability of an overload. This idea of expanding working memory capacity can be better explained in terms of expanding memory workload capacity and will be dealt with in more detail later in this chapter.

Finally we move on to the efforts to measure cognitive load directly.

3.3.4 *Has Measuring Cognitive Load Been a Means to Advancing Theory?*

CLT seemed to have been progressing well in terms of the discovery of several cognitive load effects through experiments that varied instructional design and compared performance scores. Suddenly we find Pass et al. [30] arguing 'cognitive load measurement' was central to advancing the cause of CLT. Their main point being that 'measures of cognitive load can reveal important information for CLT that is not necessarily reflected by traditional performance measures' ([30], page 1). To understand this position, we will look at how CLT went about measuring cognitive load in instructional experiments aimed to study cognitive load effects ie: in the form of subjective ratings and performance oriented measures in dual-task design (both of these will be discussed – subjective measures in this section and dual-task design later in this chapter). But first, a few words about earlier historical attempts of cognitive load measurement, as acknowledged by Sweller et al. ([23], Chap. 6).

Some Early Attempts In the early days of CLT, cognitive load was not measured directly, but was considered to be an 'assumption', based on the results of experiments examining relations between problem solving and learning. Such indirect techniques included computational models, performance during acquisition and error profiles between problems. Approaches based on *Computational models* drew parallels between human and computer problem-solving scenarios. Several experiments showed that learning strategies that required considerable problem-solving search led to inferior learning outcomes as compared to approaches that needed less problem-solving search. In computing jargon, this meant that the larger the search for a goal-state in a possibly bigger state-space required greater computational resources and perhaps increased learning time as well. Thus, learning strategies with greater problem-solving search were associated with higher cognitive load (labelled here as extraneous) and this was supported by experimental results that indicated impeded learning in the corresponding cases. Another indirect measure for CL was *performance indicators* (like instructional time and errors) *during the acquisition or learning phase*. Some experiments showed that increased cognitive load during acquisition phase negatively impacted both learning time and accuracy of the acquisition task. Another indirect measure was '*error rates*'. It was shown [31] that students often made errors at particular points when problem solving in a geometry domain, due to high working load at those points. Later Ayres [32] showed that error rates varied on mathematical tasks that required sequential calculations. Higher error rates corresponded to locations where decision making was at its greatest intensity as a result of many variables needing to be considered.

Efficiency Measures Paas et al. [33] built an efficiency measure for measuring cognitive load. Self-evaluated mental effort ratings from subjects and their test

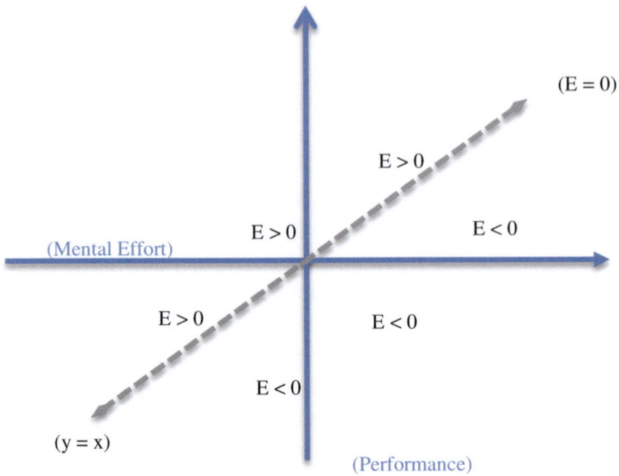

Fig. 3.7 Graphical depiction of efficiency

performance scores were used for this purpose. Efficiency here is given by the following formula,

$$E = \frac{(Z_{Ptest} - Z_{Etest})}{\sqrt{2}}$$

where E = efficiency. Standardized z-test scores of mental effort (Z_{Etest}) are subtracted from performance test scores (Z_{Ptest}) and plotted against the reference line y = x (see Fig. 3.7). Whenever $E > 0$, the instructional design is considered less efficient (and load as high) with more effort needed for comparative test performance and conversely is considered more efficient when $E < 0$ (with correspondingly less load). Theoretical concerns were raised when Hoffman and Schrow [34] objected to the meaningfulness of the quantity resulting from subtracting mental effort scores from performance scores. They proposed two refinements for this measure namely: (a) a likelihood model (based on a ratio of performance and subjective ratings), and (b) a conditional likelihood model (based on ratios of probabilities). However, all such efficiency measures, due to standardized/norm-referenced requirements, are only applicable to group performances and not suitable for accessing individual scenarios.

Subjective Measures Not satisfied with the indirect ('assumption based') approaches, a need was felt for more direct measures and researchers started looking towards learners themselves to provide a self-assessment of their subjective experience while performing various tasks. In hindsight, the effort by Paas [35] provided the much needed breakthrough. Here Pass [35] used a 9-point Likert scale ranging from "very, very low mental effort" (1) to "very, very high mental effort" (9). He argued that since learners were able to introspect the amount of mental effort they had invested during the learning and testing, their assessment of intensity of effort could be considered an 'index' of CL. Learners were asked to rate their mental effort at various points in the learning (and testing) cycle and Paas [35] found a match between self-rated mental effort and test performance. Learners who undertook a low cognitive load instructional design had superior learning outcomes and also rated their mental effort as low. These findings were quickly replicated by others with some researchers preferring to ask learners about how 'difficult' or 'easy' they found the task as opposed to asking about 'mental effort' they put into it. Ayres [36] found that subjective measures of difficulty could also detect variations for each interactive element within tasks. Furthermore, since element interactivity within tasks was associated with intrinsic load, everything seemed to be falling in place. Thereafter subjective measures of both mental effort and difficulty, continued to be a popular choice in CLT investigations – but were not without their own problems.

Inconsistencies began to surface when some studies found no significant differences between subjective measures in spite of group treatment differences on performance tests (see [37, 38]), whereas other studies reported cognitive load differences based on subjective measures but no group treatment effect on performance tests (see [37, 38]). Another case in point was a study by Kalyuga et al. [29], where each of the three experiments produced a different result: a cognitive load difference with no test effect; a cognitive load difference and a corresponding test effect as well as a lack of cognitive load difference but a test effect. Such inconsistencies were initially explained away stating statistical inevitability as 'correlation between subjective rating scales and test performances cannot be perfect' ([23], page 75). However, as more discrepancies continued to be reported, alternate strategies were also being explored. Van Gog and Paas [39] proposed that mental effort and difficulty might be distinct mental constructs leading to varied consequences. Also that, variables like the time when mental ratings were collected, personal experience, working memory capacity or even prior intelligence itself could be a source of variation. These and other similar issues continue to be investigated. Here we mention only one recent finding by Schmeck et al. [40] about differences between immediate and delayed subjective ratings. Interestingly, they investigated immediate and delayed subjective ratings, not only of mental effort and difficulty, but also from the perspective of affective variables like interest and motivation. Results showed that the delayed ratings of both mental effort and difficulty were significantly higher than the average of the six ratings made during problem solving. Problems of higher complexity seemed to be best predictors for the delayed ratings. However, the delayed ratings of affective variables like interest and motivation did not differ from the average of immediate ratings. Overall, subjective measures of cognitive load remain effective but only under qualified conditions. Let us now take a look at how physiological measures have fared for measuring cognitive load.

Physiological Measures Several attempts to measure CL have been made using physiological measures like spectral analysis of heart rate [33], cognitive papillary

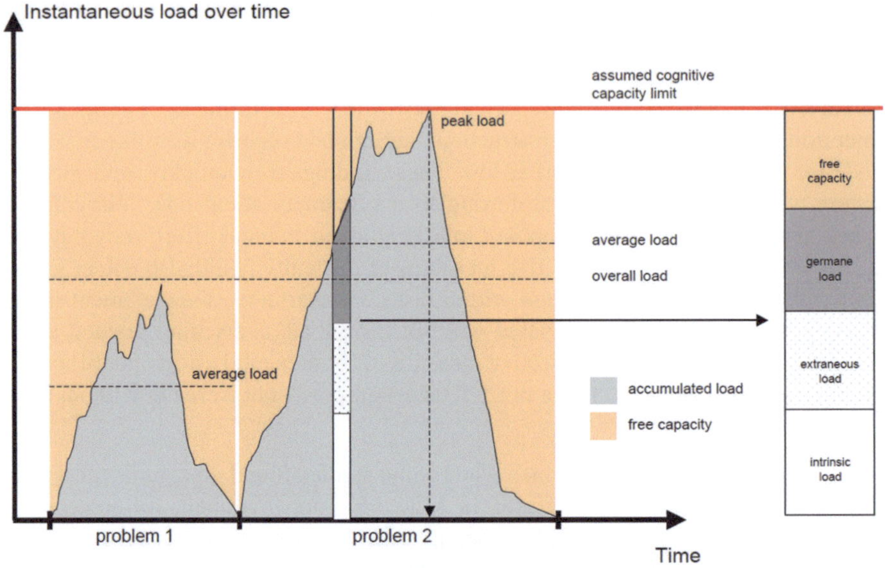

Fig. 3.8 Attributes of dynamic CL with traditional framework of CL definitions ([30], copyright © 2003 Taylor & Francis, reprinted by permission of Publisher of Taylor & Francis Ltd)

response [38] and lexical density of speech [41]. However, Xie and Salvendy [42] are a special case, as they attempted to provide a model for predicting mental workload in both single and multi-task environments. Their model was based on continuous measures (ideally from physiological sources like heart-rate or pupillary response etc.) representing instantaneous load. This instantaneous load would give rise to other values like peak load, average load, accumulated load and overall load that could then be correlated to performance measures. This was a uniquely different approach with a promising potential for real-time workload load measures.

This led Paas and colleagues [30] to attempt to explain Xie and Salvendy's [42] dynamic CL framework in terms of the traditional cognitive load types within CLT (see Fig. 3.8) and argue that a combination of performance and cognitive load measures constitute a reliable estimate of the mental efficiency of instructional methods. This was the context of the earlier comment that 'measuring CL' effectively was a means of advancing CLT. However, Sweller et al. [23] remained unconvinced, as evidenced by the following comment – '[i]n the past, physiological measures have proved insufficiently sensitive to indicate the differences in cognitive load generated by the instructional designs used by cognitive load theory' and '[i]t remains to be seen whether current attempts to find sufficiently sensitive physiological measures will prove successful'.

Even with all the success and popularity of CLT, measuring cognitive load has consistently been considered problematic [43], to the extent that Kirschner et al. [44] called it the 'holy grail' of problems within cognitive load theory. From the critical account given above it can be seen that the problem of measuring cognitive load has actually been a significant force behind the evolution of the theoretical underpinnings of CLT. Paas et al. [30], similarly had pursued these ideas over the last decade.

3.3.5 *Bridging Mental Workload and Cognitive Load Constructs*

Galy et al. [45] tested the hypothesis of an additive interaction between intrinsic, extraneous and germane cognitive load, by manipulating factors of mental workload assumed to have a specific effect on either type of cognitive load. They asserted that the study of cognitive load factors and their interaction is essential if workers' wellbeing and safety is to be improved at work. If high cognitive load requires the individual to allocate extra resources to entering information, then follows that this demand for extra resources would reduce information processing efficiency and performance. Their study tested the effects of three factors thought to act on either cognitive load type, ie task difficulty, time pressure and alertness in a working memory task. Results revealed additive effects of task difficulty and time pressure, and a modulation by alertness on behavioural, subjective and psychophysiological workload measures. Mental overload was found to be the result of a combination of task-related components, but its occurrence also depended on subject-related characteristics, including alertness. They suggested that solutions designed to reduce incidents and accidents at work should consider work organization in addition to task constraints in so far that both these factors may interfere with mental workload.

In multimedia learning, DeLeeuw and Mayer [46] made attempts to compare the three measures of cognitive load. They investigated evidence for separable measures of intrinsic, extrinsic and germane load in two college environment experiments where students were given a multimedia lesson on electric motors. Cognitive load was measured via a self-report mental effort rating instrument, response time to a secondary task and a difficulty rating scale at the end of the lesson. Correlations among the three measures were generally low. Analyses of variance indicated that the response time measure was most sensitive to manipulations of extraneous processing (created by adding redundant text), effort ratings were most sensitive to manipulations of intrinsic processing (created by sentence complexity), and difficulty ratings were most sensitive to indications of germane processing (reflected by transfer test performance). These results appear to be consistent with a triarchic theory of cognitive load in which different aspects of cognitive load may be tapped by different measures of cognitive load.

3.3.6 CLT Continues to Evolve

CLT has been proven a powerful and useful model precisely because it continues to evolve. Here we look at some problematic issues and recent developments.

One problematic issue with CLT has been identifying and measuring the different types of cognitive load. Most CLT researchers acknowledge that instructions can impose three types of CL on a learner's cognitive system. First is Intrinsic Load (IL), determined by task complexity and learner's prior knowledge. Second is called Extraneous Load (EL) that corresponds to features of the instructional design not feasible for learning. Finally, the third, Germane Load (GL) is associated with features of the instructional design that are actually beneficial for the learning. IL should be optimized in instructional design by selecting learning tasks that match learner's knowledge. EL should be minimized to reduce ineffective load and allow learners to focus on beneficial activities that impose GL. To facilitate in this process as reviewed in Chap. 2. Leppink et al. [47] developed a ten item instrument for the measurement of the three types of cognitive load using principal component analysis on data from a learning experiment. They validated this instrument with further experiments. Items one to three, related to intrinsic load, were questions about the complexity (of concepts, definitions and formula) inherent in material presented for learning. Items four to six, related to extrinsic load, were questions about the manner in which instructions were presented (eg clarity of language, effectiveness of presentation). Items seven to ten (related to germane load) inquired about the understanding of the learner (eg how much user's understanding of the topics, concepts, formula etc. had improved). Efforts like these look promising but have the obvious draw-back of being post-event and also being situation specific. This instrument was developed in a university lecture environment and worked reasonably well for statistical learning material in psychology and health sciences departments.

Difficulties in measuring and explaining general dynamics of different cognitive load types led one influential researcher, Kalyuga [48], to question the utility of arguing different types of load. Intrinsic and extrinsic load were empirically backed concepts but germane load (a later theoretical addition) could simply be argued to be an extension of intrinsic load. The idea was to decrease the bad (extrinsic) load and manage the good (intrinsic, germane) load.

Lately, CLT itself has undergone significant revision as seen in Tricot and Sweller [49] as well as Choi et al. [50]. Tricot and Sweller [49] think that 'domain-general' cognitive knowledge has frequently been used to explain acquired intellectual skill whereas the 'domain-specific' knowledge held in long-term memory may have provided a better explanation. The emphasis on domain-general knowledge could be misplaced if domain specific knowledge is the key factor driving acquired intellectual skill and '[o]nce the importance of domain-specific knowledge is accepted, instructional design theories and processes are transformed' ([49], page 1). Lending importance to domain-specific knowledge (already residing in long term memory) will also have implications for cognitive load measures, as this would involve the problem solver entering (more frequently) the state of either deep thinking or acting intuitively. Both these states have the potential to change visible behavioural patterns from that of an unsure, tinkering problem solver working with limited memory resources to that of a confident, intensely engaged problem solver.

Choi et al. [50] revised the traditional model of CLT ([51], see Fig. 3.9a) by incorporating the effects of physical environment to cognitive load and learning (see Fig. 3.9b). Physical environment considerations provide immense opportunity for multimodal measures. Consideration of the physical environment suggests that cognitive load is not a simple stimulus response concept and needs to be considered within a broader perspective of interaction.

3.4 Multimodal Interaction and Cognitive Load

This section focuses on why multimodal user interaction can be used to detect changes in a user's cognitive load. Some key advantages of this approach are error reduction and a qualified efficiency increase. We begin by introducing the issues inherent in user's action in a multimodal context and their relation to cognitive load. We then explain the dual task methodology used in experimental designs to induce load. This is followed by a discussion about analysing working memory capacity and finally how robustness is built into the proposed multimodal cognitive load measures.

3.4.1 Multimodal Interaction and Robustness

When humans interact with one another, in a natural environment, they do so in a multimodal manner. Multimodal here implies humans using multiple senses. They may use up to five of the primary senses, in parallel or serial, to make sense of a situation. However, as per Oviat [52] – 'development of computer systems has historically been a technology-driven phenomenon, with technologists believing that 'users can adapt' to whatever they build. This has resulted in reliance on instruction, training, and practice with an interface to encourage users to interact in a manner that matches a system's processing capabilities. But human-centred design advocates that a more promising and enduring approach is to model users' natural behaviour (including any constraints on their ability to attend, learn, and perform) so that interfaces are more intuitive, easier to learn and free of performance errors. The impact of this approach has been a substantial improvement in the commercial viability of next-generation systems for a wide range of real-world applications including real-time, multimodal cognitive load measures. From this we can infer that multi-modal development is a natural progression in the quest for more human-like interfaces.

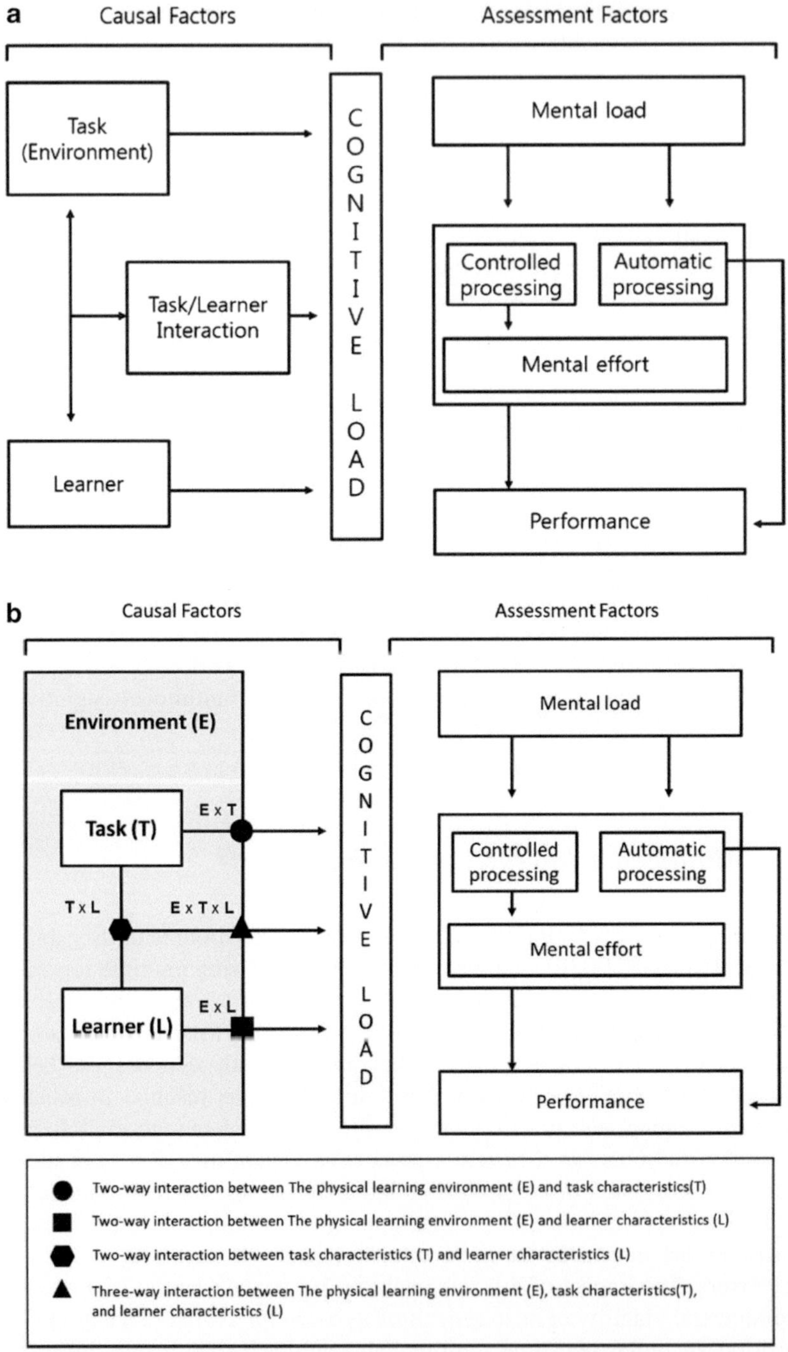

Fig. 3.9 CLT model: (**a**) Original CLT model, (**b**) revised CLT model ([50], copyright © 2014, reprinted by permission of Springer)

Nigay and Coutaz [53] had posited much earlier that, when using the term 'multimodality'; 'multi' simply refers to 'more than one' and the term 'modal' may cover the notion of 'modality' as well as that of 'mode'. In this sense, *modality* refers to the type of communication channel used to convey or acquire information. It also covers the way an idea is expressed or perceived, or the manner in which an action is performed. *Mode* refers to a state that determines the way information is interpreted to extract or convey meaning. Keeping in mind that external stimuli is experienced through sight, hearing, touch and smell, while the internal kinaesthetic stimuli may be sensed through proprioception, Turk [54] argues that any given sensing modality may be used to simultaneously estimate several useful properties of one's environment – for example, audio cues may be used to determine a speaker's identity and location, to recognize the speaker's words and interpret the prosody of the utterance, to estimate the size and other characteristics of the surrounding physical space, and to identify other characteristics of the environment and simultaneous peripheral activities. Thus one can say that, multiple sensing modalities promise to provide us with a wealth of information to support interaction both with the world and with one another.

Traditional HCI has generally been unimodal with a typical focus on windows, icons, menus & pointing (WIMP) devices. However to a certain extent, sounds from hard disk start up and blinking CPU lights have also served implicitly to provide system feedback to the user. Multimodal interfaces constitute interactive systems that seek to leverage natural human capabilities to communicate via speech, gesture, touch, facial expression, and other modalities, bringing more sophisticated pattern recognition and classification methods to human–computer interaction. The goal of research in multimodal interaction is to develop technologies, interaction methods, and interfaces that overcome existing constraints on what is possible in human–computer interaction (HCI) – towards the full use of human communication and interaction capabilities in HCI [54].

Ruiz et al. [55] listed several advantages of multimodal interfaces including robustness, naturalness, flexibility and minimising error. Several were also listed earlier by Oviatt and Cohen [56] but became less emphasized after Oviatt's [57] paper on ten myths (popular amongst the 'computationalists') of multimodal interaction became more fully understood and empirically investigated. Here we only discuss the advantages and challenges that appear relevant to MCLM.

One definite advantage of using multimodality is error reduction (or minimising error). Earlier, it was thought that, individual error-prone recognition technologies, when combined, would produce even greater unreliability (Myth #7, [57]). However, this was explained away and shown empirically to be otherwise in a multimodal system incorporating two error-prone recognition technologies of speech and handwriting recognition. Oviat demonstrated that multimodal systems actually support more robust recognition, not less—such that the error-handling problems typical of recognition technologies become more manageable. In her explanation, she argued that this increased robustness was, in part, due to leveraging from users' natural intelligence about when and how to deploy input modes effectively. In a flexible multimodal interface, people avoided using an input mode that they believed would be error-prone for certain content. Their language was also simpler which further minimized errors. When a recognition error did occur in one modality – user's conscious efforts to utilise alternate input modes appeared to resolve the error effectively. Our earlier point about language being simpler was in fact the outcome of another one of those dispelled myths, as it was generally thought that multimodal language does not differ linguistically from unimodal language (Myth #5, [57]). But it was found that multimodal language is different than traditional unimodal forms of natural language, and in many respects it is substantially simplified. One implication of this was that multimodal language is likely to be shorter, to the point and easier to process under high load or intense engagement scenarios.

One challenge for multimodal interface remains its efficient usage. This idea was also presented earlier as a myth that 'If you build a multimodal system, users will interact in a multimodal manner' (Myth #1, [57]). Whilst it is true, that users like being able to interact in a multimodal manner, there is no guarantee that they will do so necessarily. So keeping users engaged continuously in a multimodal manner is a challenge for behavioral measures of cognitive load. To be effective however, it is here that continuous physiological measures of cognitive load come into play and add to the robustness of multimodal cognitive load measures. This can be thought of as qualified efficiency, as multimodal interaction qualifies as efficient only when the human centered design successfully manages to keep the user thoroughly engaged.

Another issue to keep in mind about multimodal interaction and the monitoring of cognitive states is that multimodal signals often do not co-occur temporally in human-computer or natural human communication (Myth #3, [57]). Therefore, one cannot count on conveniently overlapped signals in order to achieve successful processing in multimodal architectures. Furthermore multimodal integration does not necessarily involve redundancy. In fact the dominant theme in user's natural organization of multimodal input is complementarity of content and not redundancy. Multimodal commands from users are not integrated in a uniform way (Myth #8, [57]). When users interact multimodally, there can be large individual differences in integration patterns. Users can adopt either a simultaneous or sequential integration pattern when combining input (eg speech and pen).

Thus the increased robustness of multimodal systems depends also a great deal on architectural designs that integrate modes synergistically. In a well-designed and optimized multimodal architecture, there will be mutual disambiguation of more than one input signals. Multimodal architecture makes parallel recognition and semantic interpretation possible, which can give a higher likelihood of the correct interpretation of input as compared to recognition based on a single input mode.

3.4.2 Cognitive Load in Human Centred Design

Due to the requirement for universal access, multimodal interfaces tend to be inherently flexible and ideal for accommodating both the changing demands encountered during mobile use and also of the large individual differences present in the user population. But cognitive load continues to play an important role in the development of multimodal interfaces in human-centred design, as actual multimodal behavioural interaction patterns appear to change with changes in load scenarios. Oviatt et al. [58] found significant changes in multimodal interaction patterns with increase in task complexity. They also found that user's task critical errors and response latencies increased systematically and significantly with increasing cognitive processing load. This demonstrates that changes in cognitive load lead to observable multimodal behavioural changes.

Adaptive human-computer interfaces are based on the core assumption that multimodal user interaction patterns will enable system to intelligently infer changes in human cognitive state. If the user starts to feel overwhelmed or under challenged, the system can adjust the pace of work via an intelligent interface to keep the user sufficiently interested and performing optimally. Furthermore, human cognitive states are not the only thing that needs to be monitored – as humans themselves are also very good at self-adaptation and maintaining an optimal state till they reach a personal threshold and after that the collapse is very rapid. This has been shown previously in the relation between load and operator performance (see Fig. 3.2) and also more recently in the subjective assessment curve (see Fig. 3.3). Therefore it is critical to monitor not just the user but also the environment and task at hand. In this sense, cognitive load theory, once again, becomes highly applicable as it incorporates all three aspects, namely germane, intrinsic and extraneous load.

3.4.3 Dual Task Methodology for Inducing Load

Subjective measures have been the most frequently used instrument to measure cognitive load. However, the traditional method of assessing working memory load has been to use a secondary task [59] in combination with a primary learning or decision task. This dual-task methodology requires a user to engage in a secondary task (eg press an indicator button whenever a fixed item changes colour) that is generally dissimilar and requires less working memory resources in addition to a primary task (eg learning/solving a class of mathematical problems). Performance in the secondary task is hypothesized to deteriorate as cognitive load increases for the primary task. Accuracy measures and response times for secondary task can then serve as reasonable indicators of changing load conditions. An obvious assumption here is the limited capacity of working memory and only when total working memory capacity is almost completely engaged by total cognitive load (see Fig. 3.5) will there be observable behavioural changes.

Dual-task approaches (in CLT) were developed and popularized by Brunken et al. [60, 61]. Methods of assessing cognitive load were classified along two dimensions, *objectivity* (subjective or objective) and *causal relation* (direct or indirect). The objectivity dimension describes whether the method used subjective self-reported data or objective observations of behaviour, physiological conditions or performance; whereas the causal relation could be direct or indirect depending on the type of relation between the observed measure and the actual attribute of interest. For example, a direct link exists between cognitive load and the difficulty of the learning materials, because this difficulty is a direct result of the intrinsic and extraneous load of the materials. An indirect link exists between navigation errors and cognitive load, as frequent errors may be caused by an incomplete mental model of the learning environment, which in turn, may itself be a result of high-cognitive load.

However, the modality of the secondary task (eg visual or auditory) is of important consideration as [61, 62] demonstrated using Baddeley's [63] assumption that audio and visual materials are processed in different subsystems of working memory. Thus an auditory modality based secondary task (eg identifying/indicating an audio tone) may not have the desired impact on a primary task that does not involve auditory processing. CLT research has made less use of dual-task methodology, mainly because secondary tasks are difficult to design (due to their primary task dependence) and may require special planning or equipment – all of which are difficult to manage in a classroom learning environment. However, the main advantage of dual-task methodology is the almost continuous measure of cognitive load during a task, whereas subjective measures are only available after completion.

Details of efforts by Brunken and colleagues to measure cognitive load can be seen in Chap. 9 in [64]. A more recent development to these approaches is the rhythm method [65]. The goal of this work was to develop a secondary task, to measure cognitive load in a direct and continuous way using intra-individual, behavioural measures. This new task was achieved by utilizing internalized cues. More specifically, a previously practiced rhythm was executed continuously by foot tapping (the secondary task) while learning (the primary task). The precision of the executed rhythm was indicative of cognitive load. The higher the precision, the lower the presumed cognitive load. These results provided evidence that rhythm precision allows for a precise and continuous measurement.

3.4.4 Workload Measurement in a Test and Evaluation Environment

In this section, we derive inspiration from the traditional recommendations for mental workload measurement in a Test and Evaluation environment (T&E) by Wierwille and Eggemeier [66] and then discuss how embedded tasks help better explain behavioural measures.

Wierwille and Eggemeier [66] reviewed empirical workload measurement techniques and recommended the following three categories, namely the (a) subjective, (b) performance-based and (c) physiological – based on their sufficient sensitivity and robustness. Typical issues that must be considered (and remain relevant to date) include sensitivity, intrusion, diagnosticity, global sensitivity, transferability and implementation requirements. Sensitivity is the degree to which a given workload technique can distinguish differences in levels of load imposed on an operator. Intrusion is the undesirable property in which introducing the workload measuring technique causes a change in operator performance. Diagnosticity is the ability to differentiate the type or cause of workload. Global sensitivity is the capability to reflect variations in different types of resource expenditure or factors that influence workload. Transferability is the capability of a technique to be used in various applications, whereas implementation requirements represent an important consideration in T&E that includes any equipment or instrumentation that is necessary to present information or record data.

Also a persistent limitation, rightly pointed out by [66], is that '[p]ure forms of operator behavior are rare in T&E. Usually operators perform functions that involve numerous forms of behavior, and the forms vary from task to task. The main objective of workload measurement during T&E must be to obtain an overall assessment. Techniques used for this purpose should have global sensitivity. Otherwise, a shift in specific forms of operator behavior may result in failure to detect workload differences'. However, of great interest to us, are the two types of secondary tasks (namely external and embedded) pointed out in the context of performance-based measures in dual-task scenarios. External tasks are the original tasks of the dual-task scenario and have little or nothing to do with the primary task. However the embedded tasks are quite different. An embedded task (eg basic communication or writing) is an operator system function that can be concurrently performed or incorporated within the task or function whose workload is to be assessed. The purpose of the embedded secondary task is essentially the same as that of an external secondary task but since it represents an actual element of system operation, the embedded secondary task does not appear artificial. However, it should have an established priority dictated by system operations, which will help minimize potential operator acceptance and intrusion problems. Several of the multimodal user behaviors (eg pen strokes, hand gestures etc.) can be interpreted as embedded secondary tasks and then a performance analysis of these tasks can be indicative of changes in cognitive load.

From the discussion above and of previous sections, we have gradually built towards the multimodal behaviour and interaction as viable indices of cognitive load. These ideas in their preliminary form were presented earlier in [67]. Much progress has taken place ever since. We now look at working memory workload capacity issues and then finally close with multimodal cognitive load measures.

3.4.5 Working Memory's Workload Capacity: Limited But Not Fixed

Clearly cognitive load on working memory is not a static value that remains fixed. On the contrary, it fluctuates, as evidenced by studies considering instantaneous load. To further complicate this issue is the assumption regarding limitation of working memory capacity. Many studies relying on the assumption of limited memory capacity also, unknowingly, presume fixed working memory *workload* capacity. This leads to erroneous inferences about performance under changing workload conditions. It is important to realize that the idea that limited working memory capacity does not necessarily imply fixed memory workload capacity as well. Humans have been known to perform much better under certain highly challenging/stressful situations or simply after repeated practice trials. Certain changes in nature of task or task environment can also motivate a user to become significantly more engaged in that task, thereby creating the impression of increased cognitive capacity. Therefore, it is relevant here to understand some basics of working memory's workload capacity for the ensuing analysis.

The number of successfully retrieved items often defines working memory *capacity* – eg the magical number seven (plus or minus two) in case of Miller [68]. However, in perceptual tasks there is another type of capacity discussed in the literature, known as *workload capacity*. Workload capacity underpins the ability to process information as processing load increases through an increase in the number of signals to be processed. Heathcote et al. [69] developed a 'gatekeeper' task and corresponding analyses that made an assessment of workload capacity in working memory. The gatekeeper task maintained information in working memory about either one or two types of attributes for the last two items studied. By allowing performance comparison in single- and dual attribute versions, the gatekeeper task provided reliable measures of working memory's workload capacity of a user. These measures, in turn, enable the understanding of individual differences, indicating where dual-task performance is better characterized by unlimited capacity and where it is better explained by fixed or limited capacity. Heathcote et al. [69] found limited capacity to be the predominant case when processing both visual and auditory attributes. Taken together, this new task and measurement approach helped the theoretical understanding of working memory's capacity and multitasking ability.

Among the more valuable tools in behavioral science is statistically fitting mathematical models of cognition to data – response time distributions in particular [70]. Study of behavioral measures, in the context of response times, reveals valuable insights. Van Zandt and Townsend [71] presented an excellent account of designs and analyses of Response Time (RT) experiments. They built upon the serial and parallel processing stages in Donder's choice reaction task [72], navigated through the speed-accuracy RT trade-off and finally presented their 'double factorial designs' for RT experiments. Double factorial design methodology separates working memory capacity from architectural issues such as serial versus parallel processing and dependence versus independence of processing channels using factorial methods and redundant targets. These methods rely on experimental designs in which stimuli vary on at least two orthogonal dimensions eg intensity and location. There are several detailed steps and intricate mathematical procedures in the implementation of this double factorial paradigm. There is also an 'sft' R package available implementing Townsend's Systems Factorial Technology [73, 74]. Analysis of mean RT is popular for the empirical evaluation of non-mathematical hypotheses of cognitive performance (eg any stimulus–response combination effects). Another way to analyze RT data is to treat them as time series. A time series is a sequence of measurements with a time index. For RT data, the index is the trial, as they themselves are measures of time. The trials may or may not occur at fixed points in time, depending on the design of the experiment. The treatment of RT data as time-series data is primarily descriptive, without focused theories to explain trends or dependencies in data. A critical component of how we understand a mental process is given by measuring the effect of varying the workload. The capacity coefficient [75, 76] is a measure of response times for quantifying changes in performance due to workload. We will be talking more about the memory workload capacity coefficient in later sections.

3.4.6 Load Effort Homeostasis (LEH) and Interpreting Cognitive Load

Multimodal cognitive load measures and the corresponding load constructs are ultimately dependent on the current theory of working memory at any given time. Most investigations regarding multimodal interface (in HCI), CLT and CTML (in educational psychology) borrowed from the popular multicomponent working memory model by Baddeley [63]. This multicomponent model originally viewed working memory as involving a central executive and two storage systems. The central executive was thought to be of limited attentional capacity, assisted by the phonological loop (based on sound and language) and the visuospatial sketchpad. Baddeley and Hitch [77] used secondary tasks to deplete the availability of short term memory in subjects performing tasks (such as reasoning or learning) that were assumed to rely on working memory. They found clear but no drastic impairment, and proposed the three-component model of working memory (see Fig. 3.10a) in place of the unitary system. Although this model of the loop explained a good range of evidence, yet it was a limited model – that failed to explain (among other things) how serial order was maintained, how the loop interacted with the long term memory and what was the biological function of phonological loop.

To address these concerns, the model was revised [79] to include an episodic buffer, assumed to be a limited capacity store that bonded together information to form integrated episodes (see Fig. 3.10b). The episodic buffer is 'attentionally' controlled by the executive and also accessible to conscious awareness. This multi-

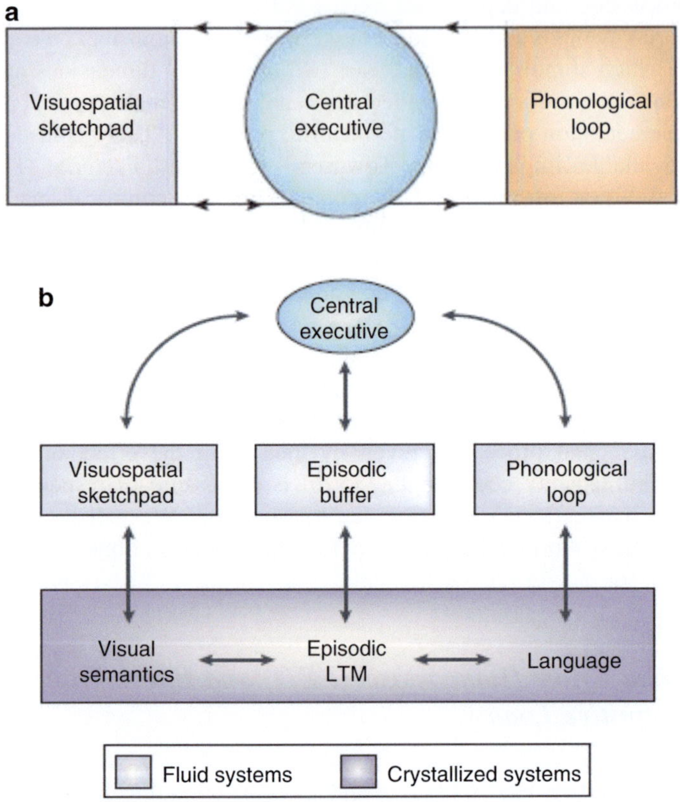

Fig. 3.10 Working memory model: (**a**) Original three component model of working memory, (**b**) Revised Multicomponent Working Memory Model (Reprinted by permission from Macmillan Publishers Ltd: Nature Reviews Neuroscience [78], copyright © 2003)

dimensional coding allowed different systems to be integrated; and conscious awareness provided a convenient tool for binding and retrieval processes. The multicomponent model has improved and continues to evolve [80], however, this is only one of the available accounts of working memory processes.

The multicomponent model, even in its most recent formulation, continues to conceive of working memory as a cognitive system separate from long term memory that is responsible for briefly maintaining and manipulating information. However, this view has been increasingly questioned (in basic working memory research) and alternates proposed that conceive of working memory as attentional processes operating on long-term memory (eg [81, 82]). Here we focus on Cowan's [83] embedded process model that has recently gained a lot of popularity. Cowan [82, 83] regarded working memory not as a separate system, but as a part of short-term memory, and extends it to include long-term memory. Incoming information is first stored in the brief sensory store. This sensory information then activates certain elements inside the long-term memory. These elements (or representations)

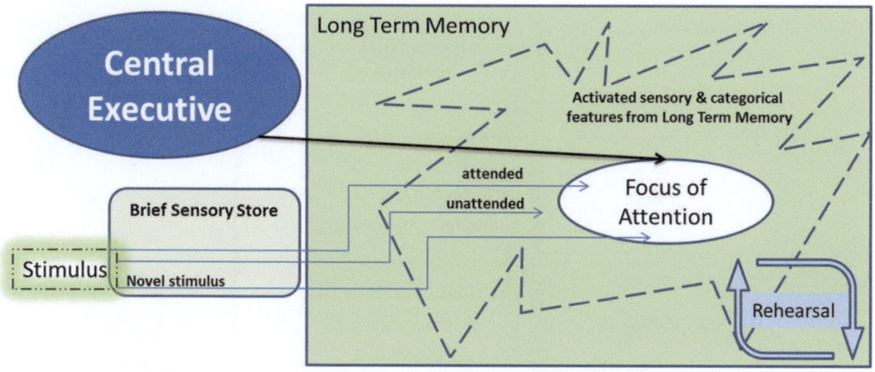

Fig. 3.11 Cowan's embedded process model (Adapted from [83])

in working memory are a subset of representations in long-term memory (see Fig. 3.11). Working memory is organized into two embedded levels. The first consists of activated long-term memory representations (shown with jagged edges in Fig. 3.11). There can be several of these and theoretically there is no limit to the activation representations in long-term memory. The second level is called the focus of attention. This focus is regarded as having a limited capacity and holds up to four of the activated representations. The activated memory consists of parts of long-term memory needed to preform (or related to) a cognitive task. Elements can be activated voluntarily or involuntarily. The amount of simultaneously active elements is still an issue of debate – but it has been shown that elements may remain active for about 10–20 s without rehearsing. Working memory holds all of these activated elements, but only about 4 ± 1 of them can be in focus at a given time and this is decided by voluntarily (or involuntarily) attention switching using the central executive. Schweppe and Rummer [84] have strongly advocated the application of embedded process models to the cognitive theory of multimedia learning. Advantages they list are: stronger focus on long-term memory, opportunity to explain and predict learning-relevant mechanisms based on attentional processes and a sophisticated concept of attention distinguishable by task and stimulus driven allocation.

So how is the conceptualization of cognitive load over different working memory models affected by the competing theories? The answer for the moment is probably not much at the subjective feedback level. But the physiological/behavioural signals can possibly reveal something that even the user is not willing to consciously acknowledge. Here we present the concept of Load-Effort Homeostasis (LEH) for individuals. LEH can be understood as the semi-conscious internal tendency of an individual to maintain performance level by adjusting personal effort level in the face of load fluctuations. This creates a distinction between the actual administered load and the perceived/acknowledged load by the user. Minor variations in load may not be fully acknowledged by the user in subjective feedback (giving rise to a subjective workload curve as argued by Estes [14] and creating and

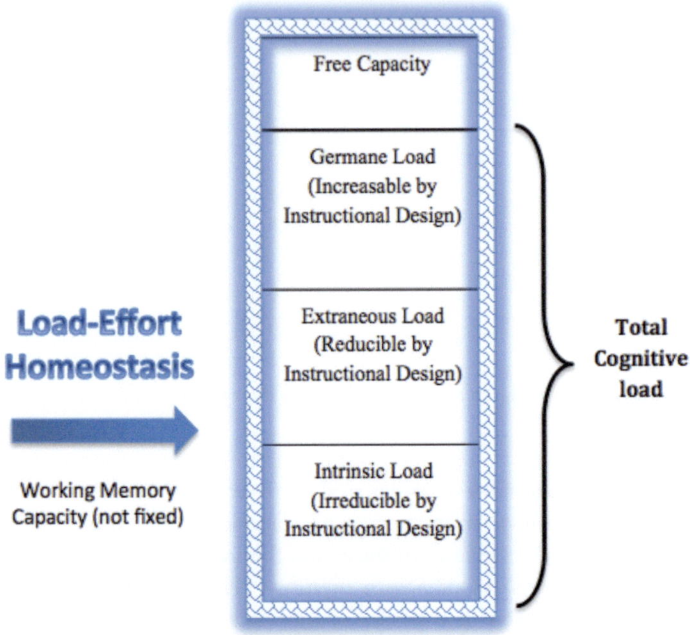

Fig. 3.12 Multicomponent WM model & LEH

observable impressions of increased working memory capacity (indicated by the blue netted pattern in Fig. 3.12 and Fig. 3.13). Several of our past research efforts have indicated that slight increases in workload (or adjacent load levels) rarely affect user subjective evaluations, though there may be observable changes in physiological and behavioural signals. A practical solution in this case is to consider five to seven levels of workload and study the effects of only the ones that are further apart and not adjacent.

LEH conceptualization has implications for the robustness of multimodal cognitive load measures. What this means is that when load level is gradually increased the user may correspondingly increase effort to maintain performance but may not think much of it and therefore such gradual/minor load changes may not be accurately reflected in offline instruments measuring load levels using subjective feedback. Furthermore, an increasingly intense user engagement can lead to an appearance of increased working memory capacity (as the same or better performance seems to result from increasing load scenarios despite same resources). This apparent increased Working Memory (WM) capacity (blue netted pattern in Fig. 3.12) interferes with the dual-task load inducing mechanism. More load can now be accommodated within available free WM capacity before any changes in performance over secondary tasks are observed. Similar reasoning can be applied to ‘focus of attention’ in embedded process model. With increasing user engagement, the span of focus of attention can increase (see Fig. 3.13) resulting in better performance or maintained

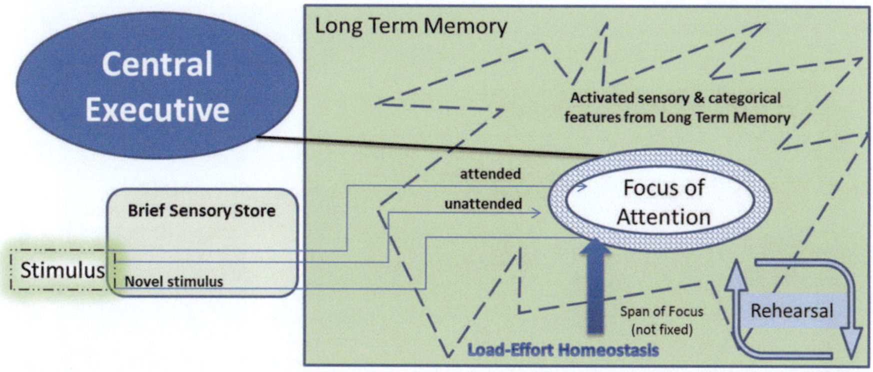

Fig. 3.13 Embedded process model & LEM

performance in the face of increasing load scenarios. However, as user effort/engagement increases, continuously monitored physiological/behavioural signals can reveal significant changes while subjective feedback shows no change.

As the theory of working memory keeps evolving, so will the framework of multimodal cognitive load measures. With these preliminary thoughts in mind, we present a model that is based on the current understanding of working memory and related empirical findings thus far.

3.5 Multimodal Cognitive Load Measures (MCLM)

In this section we bring together all the concepts that we explored earlier in this chapter and explicate the basis of MCLM. We first explain the MCLM framework and then discuss the implications of MCLM for decision-making studies and trust.

3.5.1 Framework for MCLM

The framework for multimodal cognitive load is based on five key components. The environment, user activity, a dynamic contextual filter, data-driven learning and the fusion engine (see Fig. 3.14).

The level of difficulty associated with task at hand is called the intrinsic load and (in CLT) is explained by element interactivity within task. In MCLM framework, we extend this to be covered as intrinsic factors, which, after encoding, give rise to intrinsic parameters. The importance of environment cannot be denied, as it sets the context for all user activity. Certain environments are noisy and high pressure inducing environments contributing to increased cognitive load whereas others

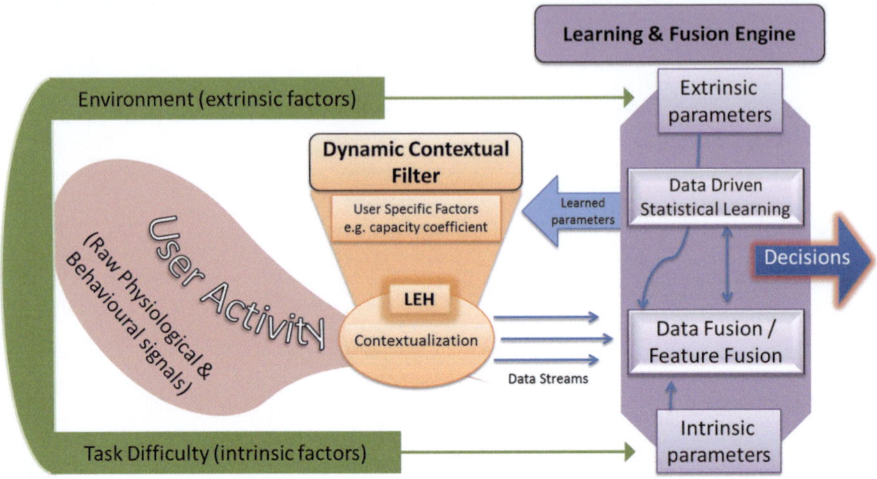

Fig. 3.14 Framework for multimodal cognitive load measures (MCLM)

can be pleasantly calming and motivating. Most recent revision of CLT [50] treats environment as a critical causal factor for cognitive load. In our case we assess environment for extrinsic factors and then encode it into extrinsic parameters. Both intrinsic and extrinsic parameters are provided to the fusion engine for establishing an interactivity platform and an expected baseline for performance.

In general, more difficult tasks should require more effort, but user *expertise* and *attitude* towards the task-at-hand generally is a stronger determinant of the actual effort applied and load subjectively experienced. Typical user expertise is encoded in long-term memory and helps provide the general threshold for a user with relevant expertise eg a lawyer with relevant law degree. However, user attitude towards a task can vary at any given moment in time. For example, a lawyer with sufficient expertise may not necessarily feel motivated enough to be engaged with the task at hand. The role of Dynamic Contextual Filter (DCF) is to provide this context for user activity signals. Each user DCF has two parts. The first part deals with relatively stable features like expertise level, experience, age, gender, capacity coefficient etc. The second part deals with the more dynamic attitudinal features of user's personality eg attention span, temperament, motivational level etc. Together, the two elements of DCF aim to provide a context for adjusting (or moderating) the psychological/behavioural signals being captured from the user in real-time. Contextualized user activity signals are fed into the learning & fusion engine. This Learning & Fusion engine has two components. One is the data driven statistical learning from continuous activity signals and this learning is then used to refine the elements of DCF. The second component is the fusion engine that uses adjusted user activity signals and parameters (extrinsic and intrinsic) to estimate decisions about load levels.

LEH is the most recent cover enhancement for DCF. First part of DCF moderates activity signals with respect to the relatively fixed expertise related user features while the second part accommodates the dynamic attitudinal related features. LEH differential contributes primarily at two levels. At higher load levels, LEH provides the context for exaggerating and giving extra importance/weight to slightest of variations in user activity signals whenever expected limits of user's cognitive capacity is being approached (and threat of failure is high). On the other hand, at lower/moderate workload levels, LEH differential provides support for accommodating/tolerating greater signal variations as users are conveniently able to match personal effort required for minor increases in workload. The behavioural/physiological signals may seem to vary but user can still reasonably manage as cognitive capacity threshold has still not been approached. Experiments are now underway to fine tune both (higher and lower load level) aspects of LEH differential.

Next we look at some implications of MCLM on cognitive modeling, decision-making and trust.

3.5.2 MCLM and Cognitive Modelling

Cognitive effort is ubiquitous. It is common to feel a cognitive task as effortful, or to decide between engaging in a demanding task and daydreaming. Cognitive effort can impact task performance in a wide variety of tasks, ranging from arithmetic to political attitude formation. It impacts economic decision-making quality as well. Cognitive effort is also fundamentally implicated in the regulation of cognitive control during goal pursuit. Despite these numerous implications, little is known about cognitive effort beyond the first principle that decision-makers seek to minimize it [85]. It is not clear why some tasks are effortful while others are not, what causes someone to withhold their effort or engage, or why we would even have a bias against effort in the first place. Likewise, subjective effort may be highly context-dependent for reasons that are not understood: In some cases, or among some individuals, cognitive effort may be sought out rather than avoided [86]. At a coarse level, "effort" refers to the degree of engagement with demanding tasks. High engagement may enhance performance by way of increasing attention. However, effort is not redundant with attention. Cognitive effort is also not motivation, though the effects of increased motivation on performance may be mediated by increased effort.

3.5.3 MCLM and Decision Making

Making decisions is one of the most complex cognitive processes and has a long history of investigation within different domain areas. For example, Morgado et al. [87] reviewed the impact of stress in decision making in the context of uncertainty and found that this complex cognitive process involves several sequential steps including analysis of internal and external states, valuation of different options available and action selection. Making good decisions implies an estimate not only of the value and the likelihood of each option but also of the costs and efforts implied in obtaining it.

3.5.4 MCLM and Trust Studies

Trust has been realized as one of the most important factors in management and organizational behavior for all personal and business decision making as well as for efficiency and task performance. It is also found to be a critical factor driving human behavior in human-machine interactions with automation systems in modern complex high-risk domains such as aviation, and the military command and control [88, 89]. For example, in organizational settings, performance improved with increased trust [90]. Various definitions are used to define trust, such as trust "*is a psychological state comprising the intention to accept vulnerability based upon positive expectations of the intentions or behavior of another*" [91] and also trust "*is to believe that the results of somebody's intended action will be appropriate from our point of view*" [92]. Different approaches are investigated to measure trust. For instance, it has been found that the extent of repeating chat abbreviations was high when communicators trusted their partners. More importantly, it was shown that there is an association between cognitive load and human's behavior which further reveals human's trust, for instance, humans who have a high level of trust in their partners use more assent and positive emotions words as a result of their satisfaction in the behaviour of their partners [93]. Therefore, trust studies can benefit from cognitive load evaluations in HCI.

3.6 Summary

In this chapter we provided the theoretical aspects of MCLM and the historical background that led to the development of its framework. From pioneering research in human factors we learned about the inability of any single measure to predict mental load and the subjective workload curve. From CLT and CTML we learned about modality principle, first-in method and environmental conditions. Viewed through the lens of HCI, we learned about error reduction and qualified efficiency in multimodality and signal fusion techniques in cognitive load measurement. Robust MCLM was investigated using dual-task experiments and the performance/response-times of secondary tasks. Behavioural measures were discussed as a special case of embedded (secondary) tasks. For robust real-time monitoring of cognitive load using multimodality, a load-effort homeostasis construct was developed and explained in the context of two WM models. Finally the framework for MCLM was explained in terms of five components namely the environment, user activity, a dynamic contextual filter and a data-driven learning and fusion engine.

References

1. N. Chater, J.B. Tenenbaum, A. Yuille, Probabilistic models of cognition: Conceptual foundations. Trends Cogn. Sci. **10**(7), 287–291 (2006)
2. T. Griffiths, C. Kemp, J. Tenenbaum, Bayesian models of cognition, in *The Cambridge Handbook of Computational Psychology*, ed. by R. Sun (Cambridge University Press, Cambridge, 2008)
3. J.L. Austerweil, S.J. Gershman, T.L. Griffiths, Structure and flexibility in Bayesian models of cognition, in *The Oxford Handbook of Computational and Mathematical Psychology*, ed. by J.T. Townsend, J.R. Busemeyer (Oxford University Press, Oxford, 2015)
4. P.D. Bruza, Z. Wang, J.R. Busemeyer, Quantum cognition: A new theoretical approach to psychology. Trends Cogn. Sci. **19**(7), 383–393 (2015)
5. N. Moray (ed.), *Mental Workload: Its Theory and Measurement* (Springer, Boston, 1979)
6. A. Westbrook, T.S. Braver, Cognitive effort: A neuroeconomic approach. Cogn. Affect. Behav. Neurosci. **15**(2), 395–415 (2015)
7. G. Johannsen, Workload and workload measurement, in *Mental Workload: Its Theory and Measurement*, ed. by N. Moray (Springer, Boston, 1979), pp. 3–11
8. W.W. Wierwille, Physiological measures of aircrew mental workload. Hum. Factors: J. Hum. Factors. Ergonomics. Soc. **21**(5), 575–593 (1979)
9. R.C. Williges, W.W. Wierwille, Behavioral measures of aircrew mental workload. Hum. Factors: J. Hum. Factors. Ergonomics. Soc. **21**(5), 549–574 (1979)
10. S. Hart, L. Staveland, Development of the NASA-TLX (Task Load Index): Results of empirical and theoretical research, in *Human Mental Workload*, ed. by P.A. Hancock, N. Meshkati, North Holland Press, Amsterdam (1988)
11. G.B. Reid, T.E. Nygren, The subjective workload assessment technique: A scaling procedure for measuring mental workload. Adv. Psychol. **52**, 185–218 (1988)
12. A. Miyake, N.P. Friedman, D.A. Rettinger, P. Shah, M. Hegarty, How are visuospatial working memory, executive functioning, and spatial abilities related? A latent-variable analysis. J. Exp. Psychol. Gen. **130**(4), 621–640 (2001)
13. C. Collet, P. Averty, A. Dittmar, Autonomic nervous system and subjective ratings of strain in air-traffic control. Appl. Ergon. **40**(1), 23–32 (2009)
14. S. Estes, The workload curve: Subjective mental workload. Hum. Factors **57**(7), 1174–1187 (2015)
15. M.S. Young, K.A. Brookhuis, C.D. Wickens, P.A. Hancock, State of science: Mental workload in ergonomics. Ergonomics **58**(1), 1–17 (2015)
16. S.G. Hart, C.D. Wickens, Cognitive workload, in *NASA Human Systems Integration Handbook* (National Aeronautics and Space Administration, Washington, DC, 2010)
17. C.D. Wickens, P.S. Tsang, Workload, in *Handbook of Human-Systems Intergration*, ed. by F. Durso, J.D. Lee, D.A. Boehm-Davis (APA, Washington, DC, 2014)
18. C.D. Wickens, A. Santamaria, A. Sebok, A computational model of task overload management and task switching, in *Proceedings of the Human Factors and Ergonomics Society Annual Meeting*, 2013, vol. 57, pp. 763–767
19. C.D. Wickens, Multiple resources and mental workload. Hum. Factors: J. Hum. Factors. Ergonomics. Soc. **50**(3), 449–455 (2008)
20. C.D. Wickens, Multiple resources and performance prediction. Theor. Issues. Ergonomics. Sci. **3**, 159–177 (2002)

21. C.D. Wickens, Multiple resource time sharing models, in *Handbook of Human Factors and Ergonomics Methods*, ed. by N.A. Stanton, A. Hedge, K. Brookhuis, E. Salas, H.W. Hendrick (CRC Press, Boca Raton, 2005), pp. 40.1–40.7
22. J.L. Plass, R. Moreno, R. Brünken, *Cognitive Load Theory*, 1st edn. (Cambridge University Press, Cambridge/NewYork, 2010)
23. J. Sweller, P. Ayres, S. Kalyuga, *Cognitive Load Theory* (Springer, New York, 2011)
24. R.E. Mayer, Multimedia learning: Are we asking the right questions? Educ. Psychol. **32**(1), 1–19 (1997)
25. R.E. Mayer, J. Heiser, S. Lonn, Cognitive constraints on multimedia learning: When presenting more material results in less understanding. J. Educ. Psychol. **93**(1), 187 (2001)
26. R.E. Mayer, Cognitive theory of multimedia learning, in *The Cambridge Handbook of Multimedia Learning*, ed. by R.E. Mayer (Cambridge University Press, New York, 2005)
27. R. Moreno, B. Park, Cognitive load theory: Historical development and relation to other theories, in *Cognitive Load Theory*, ed. by J.L. Plass, R. Moreno, R. Brünken (Cambridge University Press, Cambridge, 2010), pp. 9–28
28. R.E. Mayer, *Multimedia Learning*, 2nd edn. (Cambridge University Press, Cambridge, 2009)
29. S. Kalyuga, P. Chandler, J. Sweller, When redundant on-screen text in multimedia technical instruction can interfere with learning. Hum. Factors **46**(3), 567–581 (2004)
30. F. Paas, J.E. Tuovinen, H. Tabbers, P.W.M. Van Gerven, Cognitive load measurement as a means to advance cognitive load theory. Educ. Psychol. **38**(1), 63–71 (2003)
31. P. Ayres, J. Sweller, Locus of difficulty in multistage mathematics problems. Am. J. Psychol. **103**(2), 167–193 (1990)
32. P.L. Ayres, Systematic mathematical errors and cognitive load. Contemp. Educ. Psychol. **26** (2), 227–248 (2001)
33. F.G.W.C. Paas, J.J.G.V. Merriënboer, The efficiency of instructional conditions: An approach to combine mental effort and performance measures. Hum. Factors: J. Hum. Factors. Ergonomics. Soc. **35**(4), 737–743 (1993)
34. B. Hoffman, G. Schraw, Conceptions of efficiency: Applications in learning and problem solving. Educ. Psychol. **45**(1), 1–14 (2010)
35. F.G. Paas, Training strategies for attaining transfer of problem-solving skill in statistics: A cognitive-load approach. J. Educ. Psychol. **84**(4), 429–434 (1992)
36. P. Ayres, Using subjective measures to detect variations of intrinsic load within problems. Learn. Instr. **16**(5), 389–400 (2006)
37. B.D. Homer, J.L. Plass, L. Blake, The effects of video on cognitive load and social presence in multimedia-learning. Comput. Hum. Behav. **24**(3), 786–797 (2008)
38. P.W.M. Van Gerven, F. Paas, J.J.G. Van Merriënboer, H.G. Schmidt, Memory load and the cognitive pupillary response in aging. Psychophysiology **41**(2), 167–174 (2004)
39. T. Van Gog, F. Paas, Instructional efficiency: Revisiting the original construct in educational research. Educ. Psychol. **43**(1), 16–26 (2008)
40. A. Schmeck, M. Opfermann, T. Van Gog, F. Paas, D. Leutner, Measuring cognitive load with subjective rating scales during problem solving: Differences between immediate and delayed ratings. Instr. Sci. **43**(1), 93–114 (2015)
41. M.A. Khawaja, F. Chen, N. Marcus, Using language complexity to measure cognitive load for adaptive interaction design, in *Proceedings of International Conference on Intelligent User Interfaces (IUI 2010)*, Hong Kong, China, 2010, pp. 333–336
42. B. Xie, G. Salvendy, Prediction of mental workload in single and multiple tasks environments. Int. J. Cogn. Ergon. **4**(3), 213–242 (2000)
43. P. Ayres, F. Paas, Cognitive load theory: New directions and challenges. Appl. Cogn. Psychol. **26**(6), 827–832 (2012)
44. P. Kirschner, P. Ayres, P. Chandler, Contemporary cognitive load theory research: The good, the bad and the ugly. Comput. Hum. Behav. **27**(1), 99–105 (2011)
45. E. Galy, M. Cariou, C. Mélan, What is the relationship between mental workload factors and cognitive load types? Int. J. Psychophysiol. **83**(3), 269–275 (2012)

46. K.E. DeLeeuw, R.E. Mayer, A comparison of three measures of cognitive load: Evidence for separable measures of intrinsic, extraneous, and germane load. J. Educ. Psychol. **100**(1), 223–234 (2008)
47. J. Leppink, F. Paas, C.P.M. Van der Vleuten, T. Van Gog, J.J.G. Van Merriënboer, Development of an instrument for measuring different types of cognitive load. Behav. Res. Methods **45** (4), 1058–1072 (2013)
48. S. Kalyuga, Cognitive load theory: How many types of load does it really need? Educ. Psychol. Rev. **23**(1), 1–19 (2011)
49. A. Tricot, J. Sweller, Domain-specific knowledge and why teaching generic skills does not work. Educ. Psychol. Rev. **26**(2), 265–283 (2014)
50. H.-H. Choi, J.J.V. Merriënboer, F. Paas, Effects of the physical environment on cognitive load and learning: Towards a new model of cognitive load. Educ. Psychol. Rev. **26**(2), 225–244 (2014)
51. F.G.W.C. Paas, J.J.G.V. Merriënboer, Instructional control of cognitive load in the training of complex cognitive tasks. Educ. Psychol. Rev. **6**(4), 351–371 (1994)
52. S. Oviatt, Human-centered design meets cognitive load theory: Designing interfaces that help people think, in *Proceedings of the 14th annual ACM international conference on Multimedia (MULTIMEDIA 2006)*, 2006
53. L. Nigay, J. Coutaz, A design space for multimodal systems: Concurrent processing and data fusion, in *Proceedings of the INTERACT '93 and CHI '93 Conference on Human Factors in Computing Systems*, New York, NY, USA, 1993, pp. 172–178
54. M. Turk, Review article: Multimodal interaction: A review. Pattern Recogn. Lett. **36**, 189–195 (2014)
55. N. Ruiz, F. Chen, S. Oviatt, Multimodal input, in *Multimodal Signal Processing*, ed. by J.-P. Thiran, F. Marqués, H. Bourlard (Elsevier, London, 2010)
56. S. Oviatt, P. Cohen, Perceptual user interfaces: Multimodal interfaces that process what comes naturally. Commun. ACM **43**(3), 45–53 (2000)
57. S. Oviatt, Ten myths of multimodal interaction. Commun. ACM **42**(11), 74–81 (1999)
58. S. Oviatt, R. Coulston, .R. Lunsford, When do we interact multimodally?: Cognitive load and multimodal communication patterns, in *Proceedings of the 6th international conference on Multimodal interfaces (ICMI 2004)*, New York, USA, 2004
59. B.K. Britton, A. Tesser, Effects of prior knowledge on use of cognitive capacity in three complex cognitive tasks. J. Verbal Learn. Verbal Behav. **21**, 421–436 (1982)
60. R. Brunken, J.L. Plass, D. Leutner, Direct measurement of cognitive load in multimedia learning. Educ. Psychol. **38**(1), 53–61 (2003)
61. R. Brunken, S. Steinbacher, J.L. Plass, D. Leutner, Assessment of cognitive load in multimedia learning with dual-task methodology. Exp. Psychol. **49**, 109–119 (2002)
62. R. Brunken, J.L. Plass, D. Leutner, Assessment of cognitive load in multimedia learning with dual-task methodology: Auditory load and modality effects. Instr. Sci.: Int. J. Lear. Sci. **32**(1), 115–132 (2004)
63. A. Baddeley, *Working Memory* (Oxford University Press, Oxford, 1986)
64. R. Brünken, T. Seufert, F. Paas, Measuring cognitive load, in *Cognitive Load Theory*, ed. by J.L. Plass, R. Moreno, R. Brünken, 1st edn. (Cambridge University Press, Cambridge/New York, 2010), pp. 181–202
65. B. Park, R. Brünken, The rhythm method: A new method for measuring cognitive load—an experimental dual-task study. Appl. Cogn. Psychol. **29**(2), 232–243 (2015)
66. W.W. Wierwille, F.T. Eggemeier, Recommendations for mental workload measurement in a test and evaluation environment. Hum. Factors: J. Hum. Factors. Ergonomics. Soc. **35**(2), 263–281 (1993)
67. F. Chen, N. Ruiz, E. Choi, J. Epps, M.A. Khawaja, R. Taib, B. Yin, Y. Wang, Multimodal behavior and interaction as indicators of cognitive load. ACM Trans. Interact. Intell. Syst. **2**(4), 22:1–22:36 (2012)

68. G.A. Miller, The magical number seven, plus or minus two: Some limits on our capacity for processing information. Psychol. Rev. **63**(2), 81–97 (1956)
69. A. Heathcote, J.R. Coleman, A. Eidels, J.M. Watson, J. Houpt, D.L. Strayer, Working memory's workload capacity. Mem. Cogn. **43**(7), 973–989 (2015)
70. T. Van Zandt, How to fit a response time distribution. Psychon. Bull. Rev. **7**(3), 424–465 (2000)
71. T.V. Zandt, J.T. Townsend, Designs for and analyses of response time experiments, in *The Oxford Handbook of Quantitative Methods in Psychology: Foundations*, 1st edn. (Oxford University Press, Oxford, 2013)
72. F.C. Donders, Over de snelheid van psychische processen [On the speed of mental processes], in *Attention and performance II*, Amsterdam: North-Holland (original work published 1868), 1969, pp. 92–120
73. J.W. Houpt, J.T. Townsend, Statistical measures for workload capacity analysis. J. Math. Psychol. **56**(5), 341–355 (2012)
74. J.W. Houpt, L.M. Blaha, J.P. McIntire, P.R. Havig, J.T. Townsend, Systems factorial technology with R. Behav. Res. Methods **46**(2), 307–330 (2014)
75. J.T. Townsend, G. Nozawa, Spatio-temporal properties of elementary perception: An investigation of parallel, serial, and coactive theories. J. Math. Psychol. **39**(4), 321–359 (1995)
76. J.T. Townsend, M.J. Wenger, A theory of interactive parallel processing: New capacity measures and predictions for a response time inequality series. Psychol. Rev. **111**(4), 1003–1035 (2004)
77. A.D. Baddeley, G.J. Hitch, Working memory, in *The Psychology of Learning and Motivation: Advances in Research and Theory*, ed. by G.A. Bower, vol. 8 (Academic, New York, 1974), pp. 47–89
78. A. Baddeley, Working memory: Looking back and looking forward. Nat. Rev. Neurosci. **4** (10), 829–839 (2003)
79. A. Baddeley, The episodic buffer: A new component of working memory? Trends Cogn. Sci. **4** (11), 417–423 (2000)
80. A. Baddeley, Working memory: Theories, models, and controversies. Annu. Rev. Psychol. **63** (1), 1–29 (2012)
81. P. Barrouillet, S. Bernardin, V. Camos, Time constraints and resource sharing in adults' working memory spans. J. Exp. Psychol. Gen. **133**(1), 83–100 (2004)
82. N. Cowan, *Attention and Memory: An Integrated Framework* (Oxford University Press, Oxford, 1998)
83. N. Cowan, An embedded-processes model of working memory, in *Models of Working Memory: Mechanisms of Active Maintenance and Executive Control* (Cambridge University Press, Cambridge, 1999)
84. J. Schweppe, R. Rummer, Attention, working memory, and long-term memory in multimedia learning: An integrated perspective based on process models of working memory. Educ. Psychol. Rev. **26**(2), 285–306 (2014)
85. C.L. Hull, *Principles of Behavior* (D. Appleton-Century Company, Incorporated, New York/London, 1943)
86. J.T. Cacioppo, R.E. Petty, The need for cognition. J. Pers. Soc. Psychol. **42**(1), 116–131 (1982)
87. P. Morgado, N. Sousa, J.J. Cerqueira, The impact of stress in decision making in the context of uncertainty. J. Neurosci. Res. **93**(6), 839–847 (2015)
88. R. Mayer, J.H. Davis, F.D. Schoorman, An integrative model of organizational trust. Acad. Manag. Rev. **20**, 709–734 (1995)
89. D. Schmorrow, K. Stanney, *Augmented Cognition: A Practitioner's Guide* (Human Factors and Ergonomics Society Press, Santa Monica, 2008)
90. K.T. Dirks, D.L. Ferrin, The role of trust in organizational settings. Organ. Sci. **12**(4), 450–467 (2001)

91. D.M. Rousseau, S.B. Sitkin, R.S. Burt, C. Camerer, Not so different after all: A cross-discipline view of trust. Acad. Manag. Rev. **23**(3), 393–404 (1998)
92. B. Misztal, *Trust in Modern Societies: The Search for the Bases of Social Order* (Wiley, Oxford, 2013)
93. A. Khawaji, F. Chen, N. Marcus, J. Zhou, Trust and cooperation in text-based computer-mediated communication, in *Proceedings of the 25th Australian Computer-Human Interaction Conference: Augmentation, Application, Innovation, Collaboration*, New York, NY, USA, 2013, pp. 37–40

Part II
Physiological Measurement

Chapter 4
Eye-Based Measures

Eye-based measures of cognitive load have a number of advantages over other methods. For example, eye activity is more ubiquitous than other modalities; pupillary response and eye blink have been shown to correlate with both visual and aural cognitive tasks; eye-based data collection is less intrusive than other physiological signal data collection. Eye-based cognitive load measurement is a popular physiological measure of cognitive workload that can be used for design and evaluation of adaptive interface in various areas of human-computer interaction (HCI) research. Pupillary response has been widely accepted as a physiological index of cognitive workload. It can be reliably measured with video-based eye trackers in a non-intrusive way. With experiments of arithmetic tasks, this chapter focuses on three aspects:

- The pupillary response of cognitive load under luminance changes;
- An adaptive workload measure based on a Haar-like feature to effectively describe the dynamic characteristics of physiological response even under the effect of noisy factors;
- Boosting-based machine learning algorithm to automatically classify cognitive load.

This chapter takes an initial step to analyze the problem of robustly evaluating cognitive load based on pupillary response under complex environments (eg changes of luminance conditions), and presents a fine-grained approach for cognitive workload measurement.

4.1 Pupillary Response for Cognitive Load Measurement

Pupillary response is one type of eye activity based workload measures and it has attracted increasing attention with the appearance of video-based eye trackers in recent years [1]. The literature on the correlation between pupillary response and

F. Chen et al., *Robust Multimodal Cognitive Load Measurement*, Human–Computer Interaction Series, DOI 10.1007/978-3-319-31700-7_4

cognitive workload has been abundant over the last two decades. To study task-evoked pupillary responses, Beatty [2] conducted experiments consisting of various tasks such as language processing, reasoning, and perception. Pupil dilation was found to be a reliable indicator of processing load during the tasks. In more recent studies in [3, 4, 5], pupillary responses were measured through video-based eye tracking and the studies have shown that feature selection is critical for workload classification.

Although there is little doubt that pupillary response is a useful measure of variations in cognitive load, pupillary response is also found to be sensitive to confounding (or noisy) factors unrelated to the cognitive task, such as changes of illumination and emotional states [6–8]. Kramer [9] reported that larger changes in the pupil diameter occur in response to illumination changes than during information processing. Thus in practice comprehensive measures need to be taken to deal with such factors when using pupillary response as a workload index.

Previous studies have shown that pupillary response provides a sensitive and reliable measure for cognitive load under constrained luminance conditions. Klingner et al. [3] found subtle changes of pupil size (pupillometric measurement) induced by cognitive load variations could be detected by remote video eye tracker. In [10], pairs of subjects performed spoken dialogues during a simulated vehicle driving task. A remote eye tracker was also used to measure pupil size for estimation of the driver's cognitive load. The reliability of remote eye tracking was demonstrated by the high correspondence between pupillary response and task performance.

Pupil size can however also be affected by changes of luminance condition in the visual field – unrelated to cognitive tasks [9]. Empirical studies showed that luminance condition has a greater influence on pupil dilation compared with cognitive workload [9] [11]. Pomplun and Sunkara [12] investigated the effects and the interaction of cognitive workload and luminance changes in a gaze-controlled human-computer interaction task. There were three difficulty levels and two levels of background brightness (black and white) in the task. The experimental result showed that pupil size was significantly influenced by the factor of task difficulty and the factor of background brightness. However there was no interaction between the two factors.

Marshall [6] proposed the Index of Cognitive Activity (ICA) method to detect cognitive load under different luminance (dark and bright) conditions. The pupillary response was decomposed by wavelet transformation into detailed signals and de-noised signals. The high oscillation in the de-noised signal was then used to determine the existence of workloads. Experimental results demonstrated that ICA was effective in detecting cognitive workload. However, the classification of cognitive load has not been addressed in the work.

Xu et al. proposed a fine-grained feature extraction method for cognitive workload measurement to deal with luminance changes in pupillary response data [13]. In their work, the whole task period was divided into several intervals corresponding to different stages of the cognitive task. For each interval, normalized average pupil diameter was employed for workload measurement. It should be noted that this feature extraction strategy depends on the setting of a specific task, so it can be difficult to apply the technique to other applications involving dissimilar task types.

4.2 Cognitive Load Measurement Under Luminance Changes

In this section, arithmetic tasks are used to evaluate user cognitive load variations under luminance changes.

4.2.1 Task Design

In the arithmetic tasks [11], each subject is seated in an office environment and performs arithmetic tasks under different luminance conditions on the monitor screen. The arithmetic tasks are designed to have four difficulty levels, each with four levels of background brightness, resulting in 16 different task types in total.

In each arithmetic task, four different numbers are displayed sequentially on the center of the screen and each number is displayed for 3 s. Each subject is asked to sum up the numbers and then choose the correct answer on the screen via a mouse click. There is no time limit for the subject to choose an answer. The difficulty level of the task is determined by the range of the numbers. In the first (lowest) difficulty level, numbers range from 0 to 1 (binary number); in the second difficulty level, numbers range from 1 to 9 (1-digit number); in the third difficulty level, numbers range from 10 to 99 (2-digit number); in the fourth (highest) difficulty level, numbers range from 100 to 999 (3-digit number). At the beginning of each task, a blank screen is shown for 6 s. Then before the first number appears, different numbers of the letter "X" are displayed at the center of the screen for 3 s. The number of "X"s is the same as the number of digits in the arithmetic task. Figure 4.1 depicts the setting for each arithmetic task and a typical pupillary response [11].

In order to produce noisy luminance change signals in the pupillary response data, different levels of background luminance are shown on the screen. This is done by setting grayscale values for the four levels of background brightness (L1, L2, L3, and L4) as 32, 96, 160 and 224, respectively. Before each arithmetic task, a black background is displayed for 6 s.

Each subject has a practice attempt before the experiment (data of the practice is not recorded). The experiment starts with one-minute's rest period with black background displayed on the screen. There are two tasks for each trial type, so each subject performs 32 arithmetic tasks in total. The tasks are randomly presented during the experiment. The experiment lasts about 25 min for each subject.

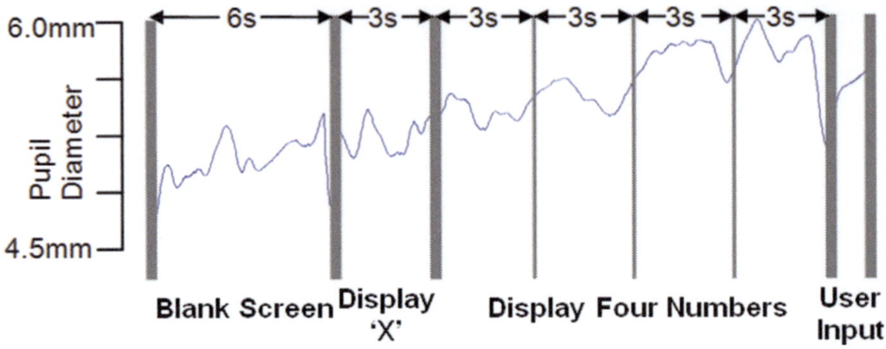

Fig. 4.1 Time setting of an arithmetic task

4.2.2 Participants and Apparatus

Thirteen 24-to-46-year-old male subjects are invited to participate in the experiment. Subjects all have normal or corrected-to-normal vision. Each subject is rewarded with a small-value gift for his participation.

A remote eye tracker (faceLAB 4.5 of Seeing Machines Ltd) is used to record the pupillary response data of each subject. The remote eye tracker continuously measures the subjects' pupil diameters at a sampling rate of 50 Hz. Visual tasks are displayed on a 21-inch monitor with a screen resolution of 1024 by 768 pixels.

4.2.3 Subjective Ratings

To validate the overall effectiveness of the cognitive workload experiment design, an ANOVA test was applied on the subjective rating scores for the four levels of task difficulty.

Figure 4.2 shows the average subjective ratings of the task difficulty levels. One-way analysis of variance (ANOVA) on the self-reporting scores showed a highly significant difference between task levels ($F_{3,48} = 108.63$, $p < 0.05$).

The response time for each task was also examined. The response time here refers to the time between *the disappearance of (or 'the removal of') the last (fourth) number* of the task and *selecting the answer*. Average response time of all subjects for each task difficulty level are shown in Fig. 4.3. It can be seen that response time has a direct relation with the task difficulty level: harder tasks take a longer response time. Results of the ANOVA test on response time of different task levels are significant ($F_{3,48} = 62.59$, $p < 0.05$). These observations about subjective rating and response time are good evidence that the designed tasks have effectively manipulated the cognitive load.

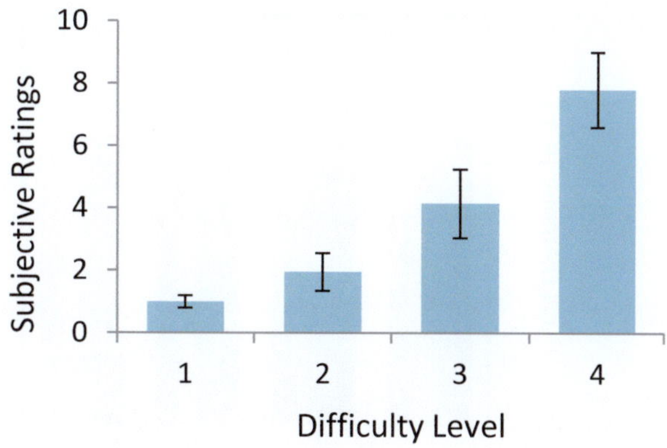

Fig. 4.2 Subjective ratings of task difficulty levels

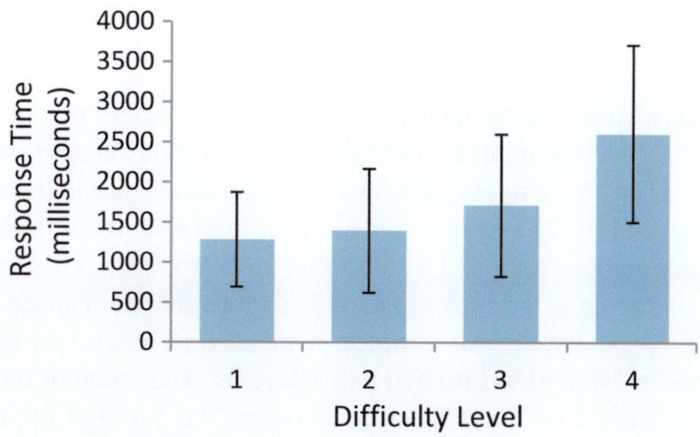

Fig. 4.3 Response time for each task difficulty level

4.3 Pupillary Response Features

In this section, pupillary response based workload measurement under changes of luminance condition is examined.

Pupillary response during eye blinks is ignored in the data preprocessing. First a coarse-grained analysis is performed to examine how background brightness and cognitive workload affect the pupillary response. The average pupil diameter is used to describe the pupillary response during the tasks. Figure 4.4 demonstrates the average pupil diameters under different task difficulty levels and background brightness conditions. Under the same difficulty level, it was seen that the pupil diameter decreases with the increase of the brightness level, and under the same brightness level, the pupil diameter has an upward trend with the rise of the difficulty level. Such observation indicates that pupil diameter is influenced by both background brightness and cognitive workload. However, the figure also reveals that luminance conditions have a greater effect size than cognitive demands in pupil diameter changes, as the pupil diameter corresponding to the highest task difficulty under the highest background brightness is smaller than that

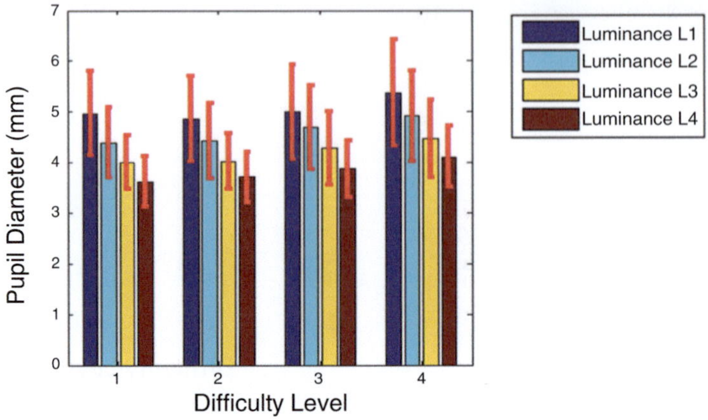

Fig. 4.4 Average pupil diameter under different task difficulty levels and background brightness conditions

corresponding to the lowest task difficulty under the lowest background brightness. This is consistent with the results of previous empirical studies [11].

The above analysis shows that the average pupil diameter is too coarse a measure to effectively assess cognitive workload with the influence of noisy factors such as luminance change. To solve this problem, a workload measure is required to both characterize cognitive load and adapt to changes caused by the confounding factors. Empirically, in this chapter, a mean-difference feature is employed by dividing each task period into two half intervals. The value of the mean-difference feature is computed as the difference between the average pupil diameter in the first half interval and that in the second half interval. To test its feasibility, distribution of the feature values extracted from different difficulty levels is shown in Fig. 4.5. It can be clearly seen that even with the interference of background luminance changes, the pupillary response based feature value rises as the task difficulty level goes up. In the corresponding ANOVA test, statistical significance ($F = 15.43$, $p < 0.05$) is obtained for the four task difficulty levels.

4.4 Workload Classification

This section discusses workload classification based on pupillary response data. The mean-difference feature used in the previous section is empirically derived through manual inspection of the experimental data. Such manual feature selection relies heavily on individual experience, and does not guarantee to be optimal, which may impair the performance of cognitive workload classification. Manual feature selection is also ineffective when the number of available features becomes large.

In order to mitigate this problem, machine learning based feature selection and classification can be used to index cognitive workload. In this section, the feature

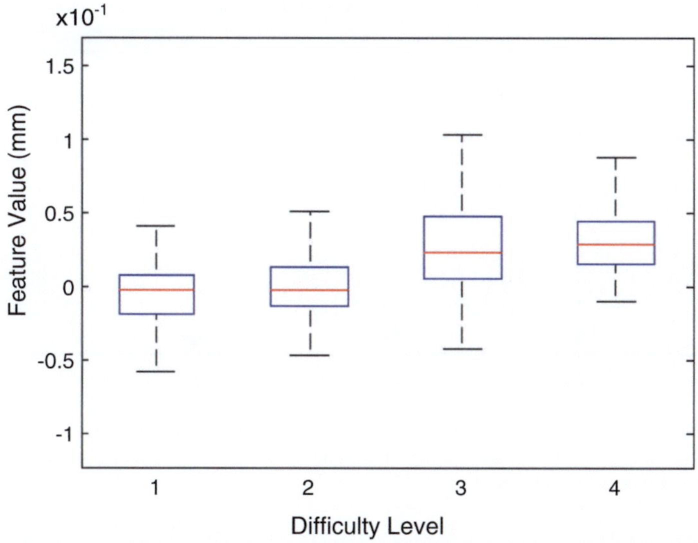

Fig. 4.5 Box plot of pupillary response based feature values (sample minimum, lower quartile, median, upper quartile, and maximum) corresponding to different task difficulty levels

space is enriched by generating a large number of dynamic characteristics which we denote as Haar-like features [14]. A boosting algorithm is applied to automatically mine the most discriminative Haar-like features for cognitive workload classification. This machine learning approach does not depend on specific task design and does not require baseline calibration, which is generally applicable to different cognitive processes. A detailed explanation of this approach is as follows.

4.4.1 Feature Generation for Workload Classification

The introduction of a Haar-like feature is inspired by the mean-difference feature described in the previous section. The mean-difference feature has been shown to be an effective measure for cognitive workload. For any physiological signal over a certain time period, the corresponding Haar-like feature could be viewed as a dynamic version of a mean-difference feature. The time period is divided into halves and the value of the corresponding Haar-like feature is the difference of the average value of the first half against that of the second half. A Haar-like feature is identified by its temporal size (the size of each half) and its centre location. The temporal size is denoted as r, the centre location as t. By varying the temporal size and the centre location, a large number of Haar-like features can be generated, which represent dynamic characteristics of physiological signals at different time-scales. Figure 4.6 shows examples of Haar-like features in pupillary response data.

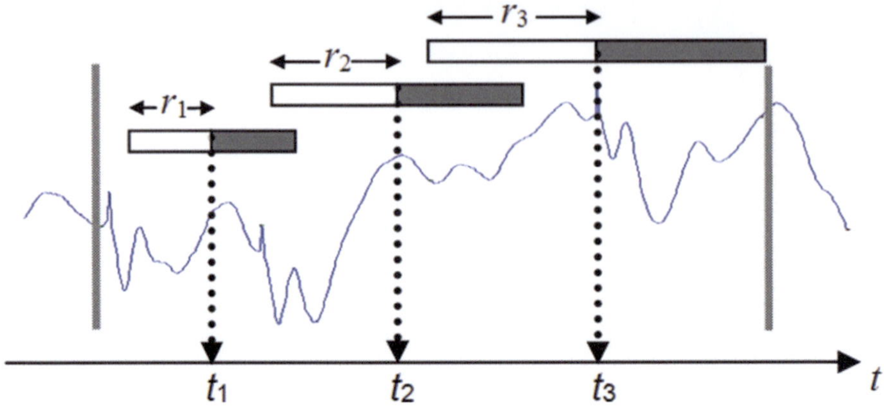

Fig. 4.6 Examples of Haar-like features. *r* represents temporal size and *t* represents the centre location of a feature. Feature value is computed as the difference between the average value of the *white* part and that of the *dark* part

4.4.2 Feature Selection and Workload Classification

The large number of Haar-like features (more than 3600 features in this work) may contain a certain amount of redundancy and some features are sensitive to noisy factors such as luminance variation or emotion arousal. To simultaneously maximize the accuracy of workload classification and minimize the impact of noisy factors, a machine learning technique is used for automatic feature selection and workload classification. The feature selection process is critical to obtaining the subset of the most useful features discriminating between different workload levels and suppressing non-relevant features. At the same time, the selected features are optimally combined to maximize the performance of workload classification. Boosting is one widely used machine learning algorithm suitable for the work in this chapter.

Boosting [15] is a classification technique which performs feature selection and classifier construction simultaneously. It is a general ensemble learning algorithm, which creates an accurate, strong classifier H by iteratively combining a number of moderately inaccurate, weak classifiers h. The definition of a strong classifier is one that has high classification accuracy on the data set, while a weak classifier is one with accuracy just larger than a random guess. The final strong classifier can be defined as:

$$H(x) = \begin{cases} 1, \text{if } \sum_{t=1}^{T} \alpha_t h_t(x) \geq 0 \\ -1, \text{otherwise} \end{cases}, \quad (4.1)$$

where α_t is a weight coefficient; h_t is a weak classifier and T is the number of weak classifiers. In a simple case, each weak classifier is attached with a feature, so the process of combining weak classifiers in Boosting is equivalent to feature selection process.

AdaBoost [16], an Adaptive version of Boosting, is employed in this section. During the process of constructing a strong classifier, AdaBoost maintains weights over the training samples; the sample weights are a reflection of the importance of training samples. Initially, sample weights are all equal. In order to select those features that are most discriminative of a given problem, in each iteration of the training process, AdaBoost selects a new weak classifier h_t with the minimal weighted classification error with respect to the training sample weight distribution, which means the newly selected weak classifier can guarantee the more important samples (samples with higher weights) are classified correctly. Then the weights of incorrectly classified samples are increased, so in the next iteration, AdaBoost can focus on these incorrectly classified samples. The process of the AdaBoost algorithm is shown in Algorithm 4.1.

ALGORITHM 4.1: AdaBoost algorithm

Input: *n* training samples $(x_1, y_1), \ldots, (x_n, y_n)$ where $y_i = -1, 1$ for negative samples and positive samples respectively. *m* candidate weak classifiers $h_1, \ldots, h_m$ where $h_i : X \to \{-1,+1\}$

Output: a Boosting classifier $H(x)$

1: Initialize sample weights: $w_{1,i} = \frac{1}{n}$, $i = 1 \ldots n$

2: **for** $t = 1 \to T$ **do**

3: Normalize the weights: $w_{t,i} \leftarrow \frac{w_{t,i}}{\sum_{j=1}^{n} w_{t,i}}$.

4: Find the classifier h_t that minimizes the error with respect to sample weights w_t:

$$h_t = \arg\min_h \varepsilon, \text{ where } \varepsilon = \sum_{i=1}^{n} \frac{1}{2} w_i \, | h(x_i) - y_i |.$$

5: The corresponding error of h_t:

$$\varepsilon_t = \sum_{i=1}^{n} \frac{1}{2} w_i \, | h_t(x_i) - y_i |.$$

6: Calculate $\alpha_t = \frac{1}{2} \ln \frac{1 - \varepsilon_t}{\varepsilon_t}$.

7: Update sample weights:

$$w_{t+1,i} = w_{t,i} \exp(-\alpha_t y_i h_t(x_i)).$$

8: **End for**

9: The final strong classifier is:

$$H(x) = \begin{cases} 1, \text{if } \sum_{t=1}^{T} \alpha_t h_t(x) \geq 0 \\ -1, \text{otherwise} \end{cases}.$$

Table 4.1 Workload classification accuracy based on pupillary response

	Two-class (%)	Four-class (%)
Training accuracy	83.6	58.9
Testing accuracy	79.3	45.0

For implementation, in the feature generation process, Haar-like features are generated from the pupillary response data. The temporal size of features ranges from 0.1 to 6 s and the centre location of features shifts over a task period with step size 0.1 s. With all the combination of different temporal sizes and centre locations, about 3600 Haar-like features were generated. Each Haar-like feature is combined with a weak classifier, so 3600 weak classifiers were obtained. The Boosting algorithm then takes the training samples and the weak classifiers as input and selects the best weak classifiers (features) and constructs a final strong classifier. It is worth noting that the whole process of feature selection requires no human input or intervention. The trained strong classifier is then used on new subject data for cognitive workload classification. Leave-One-Out Cross-Validation (LOOCV) was used to estimate the classification accuracies. Specifically, the classifier training and testing were run the same number of times as the number of subjects. At each time, the data of one subject is reserved for classifier testing and the data of the rest of subjects are used in the training. A testing accuracy was obtained each time. The final accuracy is the average accuracy of all the testing accuracies.

4.4.3 Results on Pupillary Response

Experiments were conducted on pupillary response data to evaluate the performance of the Boosting algorithm with the proposed Haar-like feature. Both two-class classification (task difficulty level 1 and 2 versus task difficulty level 3 and 4) and four-class classification were performed.

Table 4.1 shows the accuracies of two-class and four-class classification based on the pupillary response data. Boosting was seen to achieve about 80 % testing accuracy for two-class classification and about 45 % for four-class classification.

An ANOVA test was performed on the selected features for Boosting based workload classification with experiment data. Statistical significance ($F = 69$, $p < 0.05$) was obtained. Moreover, the effect size for ANOVA computed on the selected features for two-class classification is 0.58 and 0.35 for four-class classification, which demonstrates the effectiveness of Haar-like feature based Boosting.

4.5 Summary

This chapter discussed the measurement of cognitive load through eye tracking approaches and the influence of luminance conditions. The characteristics of pupillary response were investigated by analyzing the measurements acquired from different stages of the cognitive process. The experimental results demonstrated the feasibility of cognitive load measurement under complex environments using the proposed fine-grained analysis.

References

1. G.F. Wilson, R.E. Schlegel, *Operator Functional State Assessment*. NATO RTO Publication RTO-TR-HF M-104 (NATO Research and Technology Organization, Neuilly sur Seine, France, 2004)
2. J. Beatty, Task-evoked pupillary responses, processing load, and the structure of processing resources. Psychol. Bull. **91**(2), 276–292 (1982)
3. J. Klingner, R. Kumar, P. Hanrahan, Measuring the task-evoked pupillary response with a remote eye tracker, in *Proceedings of the 2008 Symposium on Eye Tracking Research & Applications*, New York, NY, USA, 2008, pp. 69–72
4. A.L. Kun, Z. Medenica, O. Palinko, P.A. Heeman, Utilizing pupil diameter to estimate cognitive load changes during human dialogue: A preliminary study, in *AutomotiveUI 2011 Adjunct Proceedings*, Salzburg, Austria, 2011
5. S. Chen, J. Epps, N. Ruiz, F. Chen, Eye activity as a measure of human mental effort in HCI, in *Proceedings of the 16th International Conference on Intelligent User Interfaces*, New York, NY, USA, 2011, pp. 315–318
6. S.P. Marshall, The index of cognitive activity: Measuring cognitive workload, in *Proceedings of the 2002 I.E. 7th Conference on Human Factors and Power Plants, 2002*, 2002, pp. 7–5–7–
7. R.F. Stanners, M. Coulter, A.W. Sweet, P. Murphy, The pupillary response as an indicator of arousal and cognition. Motiv. Emot. **3**(4), 319–340 (1979)
8. J. Xu, Y. Wang, F. Chen, H. Choi, G. Li, S. Chen, S. Hussain, Pupillary response based cognitive workload index under luminance and emotional changes, in *CHI '11 Extended Abstracts on Human Factors in Computing Systems*, New York, NY, USA, 2011, pp. 1627–1632
9. A.F. Kramer, Physiological metrics of mental workload: A review of recent progress, in *Multiple-Task Performance*, ed. by D.L. Damos (Taylor and Francis, London, 1991), pp. 279–328
10. O. Palinko, A.L. Kun, A. Shyrokov, P. Heeman, Estimating cognitive load using remote eye tracking in a driving simulator, in *Proceedings of the 2010 Symposium on Eye-Tracking Research & Applications*, New York, NY, USA, 2010, pp. 141–144
11. W. Wang, Z. Li, Y. Wang, F. Chen, Indexing cognitive workload based on pupillary response under luminance and emotional changes, in *Proceedings of the 2013 International Conference on Intelligent User Interfaces*, 2013, pp. 247–256
12. M. Pomplun, S. Sunkara, Pupil dilation as an indicator of cognitive workload in human-computer interaction, in *Proceedings of the International Conference on HCI 2003*, 2003
13. J. Xu, Y. Wang, F. Chen, E. Choi, Pupillary response based cognitive workload measurement under luminance changes, in *Human-Computer Interaction–INTERACT 2011*, 2011, pp. 178–185
14. C.K. Chui, *An Introduction to Wavelets* (Academic Press Professional, Inc., San Diego, 1992)
15. Y. Freund, Boosting a weak learning algorithm by majority. Inf. Comput. **121**(2), 256–285 (1995)
16. R.E. Schapire, Y. Freund, P. Bartlett, W.S. Lee, Boosting the margin: A new explanation for the effectiveness of voting methods. Ann. Stat. **26**(5), 1651–1686 (1998)

Chapter 5
Galvanic Skin Response-Based Measures

Galvanic Skin Response (GSR) is a measure of the conductivity of human skin, and can provide an indication of changes in the human sympathetic nervous system [1, 2]. It has recently attracted researchers' attention as a prospective physiological indicator of both cognitive load and emotional states. However, it has commonly been investigated via either a single or a few measures and typically in one experimental scenario. This chapter performs a more comprehensive study, and the highlights of the chapter include:

- GSR data captured from two different experiments are analyzed, one including text reading tasks and the other using arithmetic tasks, with each imposing multiple cognitive load levels;
- Temporal and spectral features of GSR against different task difficulty levels are examined. Obtained results show strong significance for some of the explored features, especially the spectral ones, in cognitive load measurement within the two experiments discussed;
- We assess GSR features together with eye blink features in classification of cognitive levels using support vector machines and naïve Bayesian classifiers.

5.1 Galvanic Skin Response for Cognitive Load Measurement

Among different physiological signals, GSR (also referred to as Electro-Dermal Activity (EDA)), is comparatively low-cost and easily-captured. The method involves measuring the electrical conductance of the skin through one or two sensor(s) usually attached to some part of the hand or foot. Skin conductivity varies with changes in skin moisture level (sweating) and can reveal changes in the sympathetic nervous system.

F. Chen et al., *Robust Multimodal Cognitive Load Measurement*, Human–Computer Interaction Series, DOI 10.1007/978-3-319-31700-7_5

GSR has recently been investigated in relation to mental status and emotions. Nakasone et al. [3] successfully used skin conductance and muscle activity for emotion detection. In another study, skin conductance was measured to differentiate between a stress condition and a cognitive load condition, seeking the ability to detect stress states [4]. Shi et al. [5] also assessed GSR in stress and cognitive load situations and found correlations between readings of this signal and cognitive load. Engstroem et al. [6] found a weak effect of cognitive load on physiological signals including mean skin conductance. Ikehara and Crosby [7] evaluated GSR in relation to two levels of cognitive load. In contrast with other studies, they found skin conductance increased as task difficulty increased and explained it as a result of the easy task being tedious and too simple. Wilson [8] analysed several physiological measures during different stages of flight in a simulator and found an increase in EDA response during take-off and landing which were the procedures expected to place the most cognitive demands on pilots. Haapalainen et al. [9] assessed the mean, variance and median of GSR and other physiological signals in relation to two cognitive load levels. They did not obtain any satisfactory results for GSR and explained that it might be related to the type of tasks or that their GSR sensors might not have been sensitive enough.

Support Vector Machines (SVM) and Naïve Bayesian classifiers are two popular machine learning algorithms in HCI studies [10]. Some previous works have used SVM for recognizing drowsiness [11] or different emotions [4, 12, 13] via physiological features. Naïve Bayesian classifiers have been used for detecting human emotions from facial expressions [14] or physiological signals [12]. In this chapter, these two types of classifiers are used for cognitive load classification of GSR and blink features.

Two experiments of arithmetic tasks and reading tasks are conducted in this chapter to demonstrate the use of GSR in cognitive load measurement.

5.2 Cognitive Load Measurement in Arithmetic Tasks

5.2.1 Task Design

Task Description This experiment included 8 arithmetic tasks with 4 difficulty levels. Each subject performed two trials of each task difficulty level and the whole eight trials were performed in a randomised order. The first to fourth difficulty levels involved binary numbers (0 and 1), one-, two- and three-digit numbers respectively. In each task, four numbers were shown one by one, each for 3 s. Subjects were asked to add these four numbers together and select the correct answer from three numbers which were presented at the conclusion of each trial. Trials with the same difficulty level included different numbers. Before displaying the first number of each task, one to three "X" symbols (according to the number of digits in the task) were presented for 3 s. There was no time limit for answer input [1, 2].

Apparatus To collect galvanic skin response, a GSR device from ProComp Infiniti of Thought Technology Ltd was used and the sensors were attached to the subjects' left hand index and ring fingers. The sampling frequency was 10 Hz. Eye activity data was also recorded with a remote eye tracker (faceLAB 4.5 of Seeing Machines Ltd) which operated at a sampling rate of 50 Hz and continuously recorded eye data. A 21-inch LCD monitor and a common computer mouse were peripherals for interaction between participants and a PC running the tasks. Another PC collected the signals through GSR sensors and was synchronised with the machine displaying the experimental stimuli.

Participants Thirteen 24 to 35-year-old male volunteers participated in this experiment. They signed a consent form before the experiment and were awarded with a $10 movie voucher for their participation. The experiment was approved by Human Research Ethics Committee of the University.

5.2.2 *GSR Feature Extraction*

It was observed that the GSR values are highly varied. They differ from person to person. When the recorded values are analysed without any pre-processing, no significant results can be obtained. It was hypothesised that some sort of normalisation which counteracts this variation between subjects may be helpful [1]. As will be discussed in the following sections, the data of all subjects were scaled in the same range by using a normalization procedure and this hypothesis was confirmed by the results. Time and frequency domain features were analysed, and each feature was averaged between tasks with the same difficulty levels for each subject. A one-way ANOVA test was applied to statistically evaluate each temporal/spectral feature [1], [2]. Figure 5.1 shows the GSR signal of one of the subjects.

5.2.2.1 Time Domain Features

The summation of GSR values (accumulative GSR) was calculated from the appearance of the first number to the input of the answer. This period is referred to as the task time. In addition, the GSR value of each task was averaged over task time. In order to omit the subjective differences, each participant's data was normalised by dividing the task signal by the mean value of all tasks of the subject:

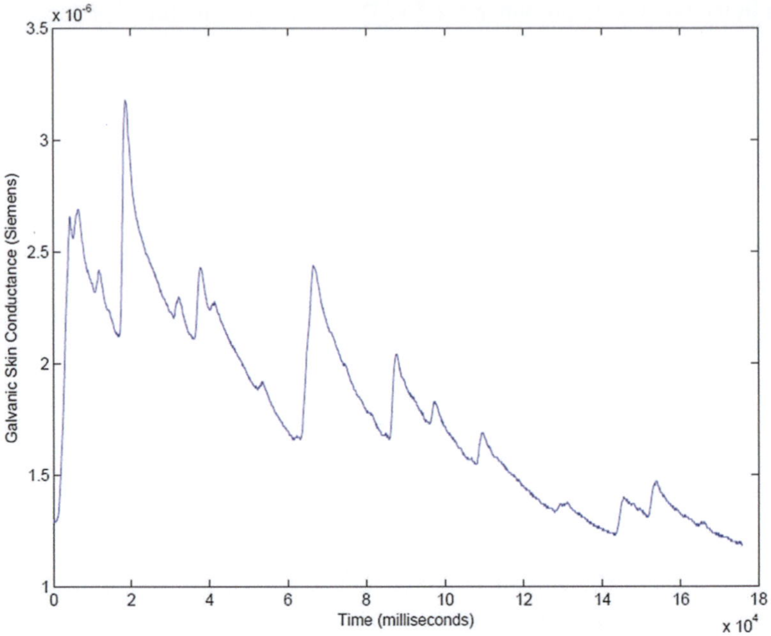

Fig. 5.1 GSR signal of subject 1

$$\text{Normalised\_GSR(i, k, t)} = \frac{GSR(i,k,t)}{\frac{1}{m}\sum_{j=1}^{m}\sum_{t=1}^{Tij} GSR(i,j,t)} \tag{5.1}$$

where GSR(i,k,t) is the value of each data-point at time t of task k of subject i, and m is the number of tasks (in this experiment m = 8). Accumulative GSR and average GSR were calculated for task k of subject i as following:

$$\text{Acc\_GSR(i, k)} = \sum_t Normalised\_GSR(i,k,t) \tag{5.2}$$

$$\text{Avg\_GSR(i, k)} = \frac{\sum_t Normalised\_GSR(i,k,t)}{T} \tag{5.3}$$

where T is the whole task time and t is a sampling time.

5.2.2.2 Frequency Domain Features

In frequency domain, different variations of the power spectrum are examined:

$$\text{P}(\omega) = \frac{1}{N}\text{Y}(\omega)Y^*(\omega) \tag{5.4}$$

where P is power spectrum, ω is angular frequency, N is the length of signal, Y is frequency spectrum and Y* is the complex conjugate of Y. The power feature was observed to have non-zero values in frequencies less than 1 Hz, mostly less than 0.5 Hz.

For each subject, the power spectrum of each task (both as a whole and cut into frames) was calculated. For the whole tasks, each task was normalised by the mean of all tasks of the subject. In the segmentation, each task was divided into frames of 16, 32 and 64 data-points length, the power spectrum was calculated over each frame. Each task was normalised by each frame k of any task j of each subject i by dividing the average power value of the frame (power_frame(*i,j,k*)) by average power of all frames of the particular subject:

$$\text{N\_power\_frame}(i, j, k) = \frac{power\_frame(i,j,k)}{\frac{1}{m}\sum_{l=1}^{m} power\_frame(i,j,l)} \tag{5.5}$$

where m is the total number of frames of subject i. The whole procedure was performed four times for each task of each subject, once on the whole task's data and once for every segmentation. In each run, power features of direct current (DC, frequency = 0), the whole frequency range (excluding DC part), and below 1 Hz (as this range had the most non-zero power values) were calculated.

5.2.3 *Feature Analyses*

Time Domain As mentioned before, GSR values are subject-dependent, ie different people have different GSR value levels. This can be observed from Fig. 5.2a. Although there is an increasing trend between task difficulty and average sum feature, the huge standard deviations reveal the magnitude of the between-subjects difference for each task. The results of statistical analysis (ANOVA test) of the accumulative GSR without any normalisations were insignificant ($p = 0.9162$). Data of all subjects were scaled in a similar range and the normalised data were analysed. The effectiveness of this normalisation can be observed from Fig. 5.2b: the increasing trend of the feature versus task difficulty levels exists and the value for each difficulty level has a much smaller standard deviation.

This normalisation leads to highly significant results in respect to differentiating between the four task difficulty levels ($F = 7.21$, $p < 0.001$). It can be observed that accumulative GSR produced results which were much more significant than those produced by average GSR (with normalisation: $F = 3.31$, $p = 0.0279$; without normalisation: $F = 0.03$, $p = 0.9940$).

Frequency Domain In the frequency domain, as in the time domain, without normalisation no significant results were obtained. Tables 5.1 and 5.2 show the results of statistical analysis of subjective-normalised frequency domain features (NDP: number of data-points in each segment).

It can be seen from Table 5.1 that the whole task and all the segmentations produce significant results. However, the best result is obtained from using the

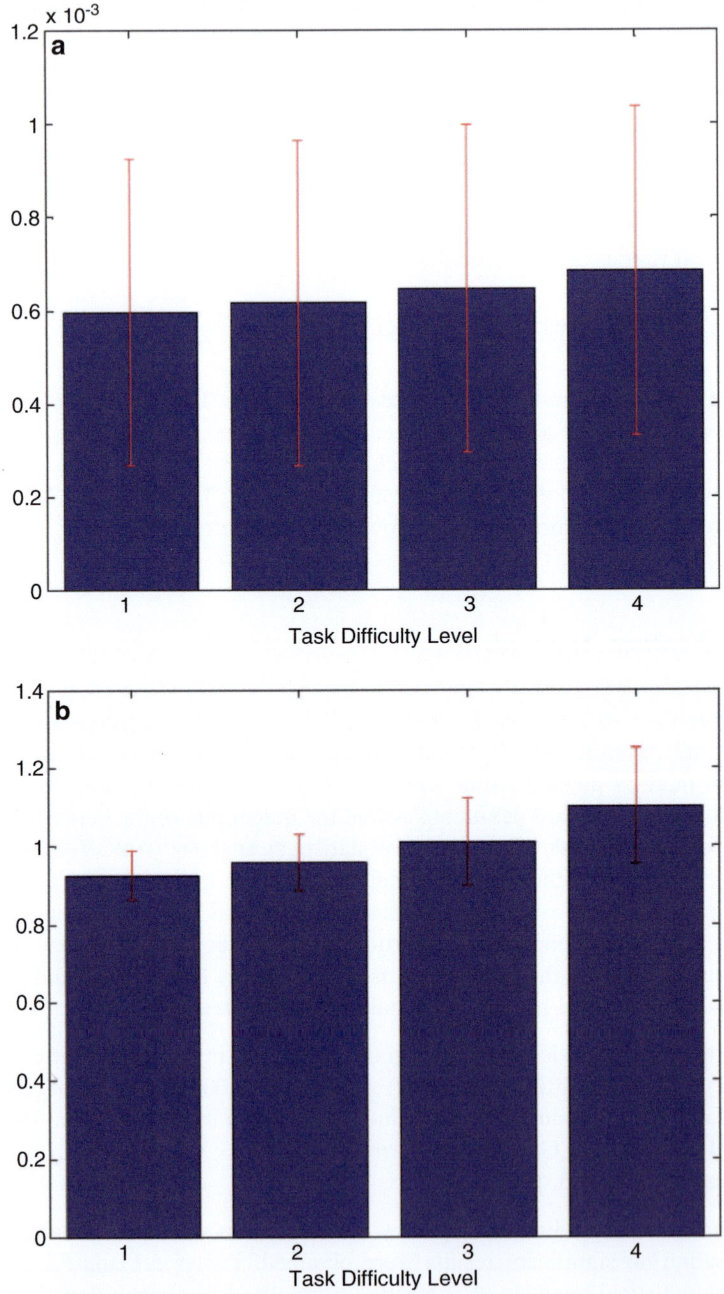

Fig. 5.2 Mean and standard deviation of accumulated GSR values for all subjects: (**a**) without normalisation, (**b**) with normalisation

Table 5.1 ANOVA test results of average GSR frequency power of DC part in the experiment

Whole task	NDP = 16	NDP = 32	NDP = 64
F = 4.96	F = 3.22	F = 3.14	F = 3.08
p = 0.0047	p = 0.0316	p = 0.0346	p = 0.0371

Table 5.2 ANOVA test results of average GSR frequency power (excluding DC) in the experiment

Whole task	NDP = 16	NDP = 32	NDP = 64
F = 3.11	F = 5.36	F = 5.14	F = 4.65
p = 0.0358	p = 0.0031	p = 0.0039	p = 0.0066

whole task. On the other hand, a slightly different pattern is observed regarding segmentations. Although these results are very close, smaller frames result in a little more significance.

Table 5.2 shows that the smaller the frame length, the more significant the results are. The best results, which are highly significant, belong to 16-datapoint-length frames. The repeated results as shown in Table 5.2 were almost achieved when analyzing the frequencies below 1 Hz. This can be explained with the observation mentioned before: the most non-zero values of the signals relate to frequencies below 1 Hz.

5.3 Cognitive Load Measurement in Reading Tasks

5.3.1 Task Design

Task Description Three silent reading tasks were performed. Each task consisted of four slides of text with each slide presented for 30 s. The participants were supposed to find words of certain lengths in each slide. There were three task difficulty levels: the easiest task required finding three-letter words; in the medium task subjects should find three- and four-letter words; in the hardest task subjects were supposed to find three-, four- and five-letter words. They were asked to click on the left, middle or right mouse button when finding a three-, four- or five letter word respectively. Task difficulty orders were randomly assigned.

Apparatus The galvanic skin response was recorded using a GSR device from ProComp Infiniti of Thought Technology Ltd and the sensors were attached to the subjects' left hand index and ring fingers. The sampling frequency was 10 Hz. A 19-inch LCD monitor and a standard computer mouse were peripherals for interaction between participants and a PC running the tasks. Another PC collected the signals through GSR sensors and was synchronised with the first one.

Participants Twelve 16 to 46-year-old male and female volunteers participated in the experiment. Each participant was awarded with a chocolate bar as an incentive toparticipate. Before performing the experiment, each subject signed a form allowing the experimenters to collect and use the data for research. The experiment was approved by Human Research Ethics Committee of the University.

5.3.2 GSR Feature Extraction

In this experiment, analysis similar to the data analysis in Sect. 5.2 was performed. The features were calculated using the same formulas. However, there were some differences. As mentioned before, the reading experiment consisted of three tasks of different difficulty levels for each participant, every task level was performed once by each subject, and the duration of the tasks were longer than the arithmetic tasks. The time and frequency domain features were analyzed and a one-way ANOVA test was applied to statistically evaluate each time/frequency domain feature [1, 2].

Time Domain Features The accumulative GSR during task time was calculated and the GSR value of each task over task time was also averaged. A similar normalisation was performed: dividing data of each task of each subject by the mean of all tasks of that subject.

Frequency Domain Features For each subject, spectral features of each task were calculated. The analysis was performed for the whole tasks and different segmentations. For the whole tasks, each power spectrum was normalised by average frequency power of all tasks of the subject. In segmentation, each task was divided into frames of 16, 32, 64 and 128 data-points length. The power spectrum over each frame was calculated and each frame of any task was normalised by dividing the power value of that frame by the average of all frames of all tasks of that particular subject. The average frequency power was examined for DC part and the whole frequency range excluding DC.

5.3.3 Feature Analyses

Time Domain As for the experiment in Sect. 5.2, due to the subjective nature of GSR values, temporal features of this experiment did not produce significant results before being normalised ($F = 0.24$, $p = 0.7868$ for accumulative GSR; $F = 0.09$, $p = 0.9676$ for average GSR). The normalisation produced significant results for average GSR ($F = 2.98$, $p = 0.0444$) and improved accumulative GSR results ($F = 1.66$, $p = 0.1925$).

Table 5.3 ANOVA test results for power feature of DC parts in the experiment

Whole task	NDP = 128	NDP = 64	NDP = 32	NDP = 16
F = 8.43	F = 3.7	F = 3.51	F = 3.52	F = 3.43
p = 0.0002	p = 0.0203	p = 0.0249	p = 0.0246	p = 0.0269

Table 5.4 ANOVA test results for power feature (excluding DC) in the experiment

Whole task	NDP = 128	NDP = 64	NDP = 32	NDP = 16
F = 0.15	F = 4.76	F = 3.98	F = 2.19	F = 1.54
p = 0.8652	p = 0.0067	p = 0.0152	p = 0.1059	p = 0.2200

Frequency Domain Due to the variable nature of GSR values between subjects, no significant results were obtained before normalisation was applied. Tables 5.3 and 5.4 show the results of statistical analysis of normalised spectral features. The whole tasks led to the most significant results when assessing the frequency spectrums in DC part of the signal, and the segmented signals produced better results when the DC part is excluded. This is consistent with the findings of the experiment in Sect. 5.2.

5.4 Cognitive Load Classification in Arithmetic Tasks

5.4.1 Features for Workload Classification

In this section, eye-based features derived from eye activity data recorded in the experiment of arithmetic tasks described above are incorporated together with GSR features for the cognitive load classifications. Two GSR and two blink features were calculated for each task:

- Cumulative GSR (summation of GSR values over task time)
- GSR power spectrum (frequency power)
- Blink number (number of blinks recorded during the task)
- Blink rate (number of blinks recorded during the task divided by task time)

The time between the appearance of the first number and input of the answer was considered as the task time in which every feature was computed. It was observed that GSR and blink values differ from person to person. In order to omit the inter-participant differences, each feature of task j of participant i was calibrated by dividing it by the average of all similar features of all tasks of that subject. Furthermore, each feature between tasks was averaged with the same difficulty levels for each subject. Figures 5.3 and 5.4 show the average values of the studied (calibrated) features in the four task levels.

A one-way ANOVA test was applied to statistically evaluate cognitive load level discrimination of each feature. Table 5.5 represents the results of statistical analysis

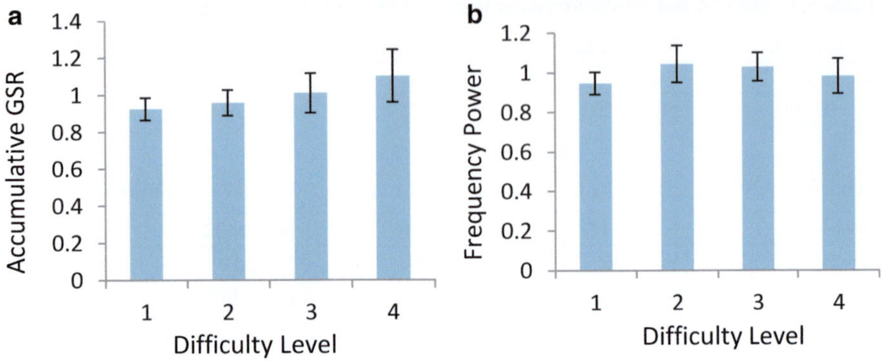

Fig. 5.3 Average GSR features of all subjects for the four task levels: (**a**) accumulative GSR, (**b**) GSR frequency power

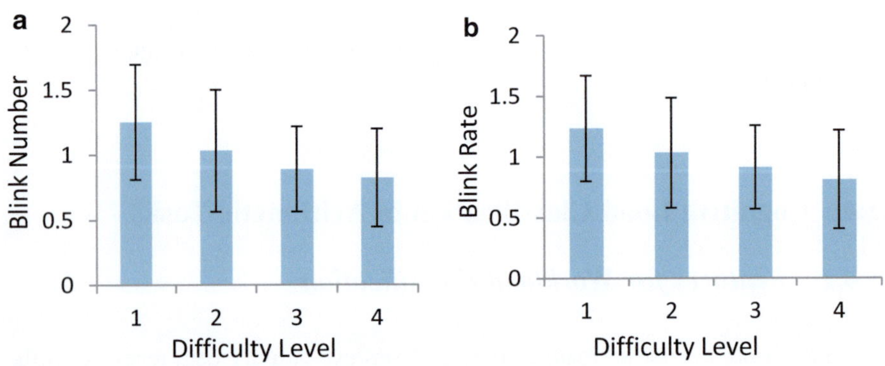

Fig. 5.4 Average blink features of all subjects for the four task levels: (**a**) blink number, (**b**) blink rate

(ANOVA test) of the studied features for four task difficulty levels. As can be seen, the results are significant for all four features.

5.4.2 *Classification Results*

In this study, SVM and Naïve Bayes classifiers were applied for cognitive load classifications. For every feature, two- and four-class classifications were examined. In the two-class classification, difficulty levels of one and two were considered as low load, and difficulty levels of three and four were considered as high load. The cross validation method was leave-one-subject-out. In other words, in each round the classifier was trained by the data of all subjects except one and data

Table 5.5 ANOVA results of features for four task difficulty levels

Feature	Results
Accummulative GSR	$F_{3,48} = 7.22$, $p < 0.05$
GSR frequency power	$F_{3,48} = 4.07$, $p < 0.05$
Blink number	$F_{3,48} = 3.37$, $p < 0.05$
Blink rate	$F_{3,48} = 3.22$, $p < 0.05$

of the remaining subject was used for testing. The classification accuracies of all rounds were averaged.

Tables 5.6, 5.7, 5.8, and 5.9 show the cognitive load classification accuracies of the single features for two and four classes. Results of all features are reasonable, GSR features outperform blink features in two-class classification and results of blink number are better than those of blink rate. It is also worth mentioning that in most cases the two types of classifiers had very close or even similar (Table 5.8) performances on the single features. The largest difference is in classifying by use of accumulative GSR (Table 5.6) where classification accuracies of Naïve Bayes learners are about 5 % higher than those of SVM in both 2-class and 4-class classifications.

In the next step, cognitive load classification was attempted using a combination of GSR and blink features. The combination of blink number and GSR frequency power resulted in the highest classification accuracies which can be seen in Table 5.10. Comparison with Tables 5.7 and 5.8 reveals that combining blink number and GSR frequency power improves the classification accuracy in both two- and four-class classifications up to about 10 % for the former and 16 % for the latter. It can be observed that for a combination of the two modalities (Table 5.10), similar to single feature cognitive load classifications, the classification accuracies of SVM and Naïve Bayes classifiers are close.

5.5 Summary

GSR is a nonintrusive easily-captured physiological signal which is being explored as a method of measuring cognitive load. In this chapter, different time and frequency-domain features of GSR were investigated in conjunction with multiple difficulty levels of arithmetic and reading experiments. Normalisation was applied to overcome the subject-dependent variance of the GSR data. The results showed that normalisation effectively improves the ability to detect differences in the cognitive load levels via mean and cumulative GSR as well as spectral features. In the frequency domain, the behaviour of the power spectrum of the whole tasks as well as smaller segments within each task were investigated, within the DC range as well as the rest of the frequency range. Within the DC range, the best results were associated with the whole tasks; however, when analysing frequencies below 5 Hz or below 1 Hz, segmented signals produced better cliassification in both experiments.

Table 5.6 Classification accuracies of accumulative GSR

Classification algorithm	2-Class (%)	4-Class (%)
SVM	66.4	34.6
Naïve Bayes	71.2	40.4

Table 5.7 Classification accuracies of GSR frequency power

Classification algorithm	2-Class (%)	4-Class (%)
SVM	66.4	37.5
Naïve Bayes	65.4	35.6

Table 5.8 Classification accuracies of blink number

Classification algorithm	2-Class (%)	4-Class (%)
SVM	62.5	40.0
Naïve Bayes	62.5	40.0

Table 5.9 Classification accuracies of blink rate

Classification algorithm	2-Class (%)	4-Class (%)
SVM	57.5	31.3
Naïve Bayes	55.0	32.5

Table 5.10 Classification accuracies of blink number + GSR frequency power

Classification algorithm	2-Class (%)	4-Class (%)
SVM	71.5	53.6
Naïve Bayes	75.0	50.0

This chapter applied classification algorithms to blink and GSR features and combinations of them. Accumulative GSR, power spectrum of GSR, blink number and blink rate were significantly distinctive and had reasonable accuracies in both two- and four-class classifications of cognitive load using support vector machines and Naïve Bayes classifiers. Combining GSR and blink features improved the classification accuracy.

References

1. N. Nourbakhsh, Y. Wang, F. Chen, GSR and blink features for cognitive load classification, in *Human-Computer Interaction – INTERACT 2013*, ed. by P. Kotzé, G. Marsden, G. Lindgaard, J. Wesson, M. Winckler (Springer, Berlin/Heidelberg, 2013), pp. 159–166
2. N. Nourbakhsh, Y. Wang, F. Chen, R.A. Calvo, Using galvanic skin response for cognitive load measurement in arithmetic and reading tasks, in *Proceedings of the 24th Australian Computer-Human Interaction Conference*, New York, NY, USA, 2012, pp. 420–423
3. A. Nakasone, H. Prendinger, M. Ishizuka, Emotion recognition from electromyography and skin conductance, in *Proceedings of The Fifth International Workshop on Biosignal Interpretation (BSI-05)*, 2005, pp. 219–222
4. C. Setz, B. Arnrich, J. Schumm, R. La Marca, G. Tröster, U. Ehlert, Discriminating stress from cognitive load using a wearable EDA device. IEEE Trans. Inf. Technol. Biomed. **14**(2), 410–417 (2010)

5. Y. Shi, N. Ruiz, R. Taib, E. Choi, F. Chen, Galvanic Skin Response (GSR) as an index of cognitive load, in *CHI '07 Extended Abstracts on Human Factors in Computing Systems*, ed. by R. Mary Beth, pp. 2651–2656
6. J. Engström, E. Johansson, J. Östlund, Effects of visual and cognitive load in real and simulated motorway driving. Transport. Res. F: Traffic Psychol. Behav. **8**(2), 97–120 (2005)
7. C.S. Ikehara, M.E. Crosby, Assessing cognitive load with physiological sensors, in *Proceedings of the 38th Annual Hawaii International Conference on System Sciences (HICSS 2005)*, Hawaii, USA, 2005
8. G.F. Wilson, An analysis of mental workload in pilots during flight using multiple psychophysiological measures. Int. J. Aviat. Psychol. **12**(1), 3–18 (2002)
9. E. Haapalainen, S. Kim, J.F. Forlizzi, A.K. Dey, Psycho-physiological measures for assessing cognitive load, in *Proceedings of the 12th ACM International Conference on Ubiquitous Computing*, New York, NY, USA, 2010, pp. 301–310
10. F. Lotte, M. Congedo, A. Lécuyer, F. Lamarche, B. Arnaldi, A review of classification algorithms for EEG-based brain-computer interfaces. J. Neural Eng. **4**(2), R1–R13 (2007)
11. M.V.M. Yeo, X. Li, K. Shen, E.P.V. Wilder-Smith, Can SVM be used for automatic EEG detection of drowsiness during car driving? Saf. Sci. **47**(1), 115–124 (2009)
12. R. Horlings, D. Datcu, L.J.M. Rothkrantz, Emotion recognition using brain activity, in *Proceedings of the 9th International Conference on Computer Systems and Technologies and Workshop for PhD Students in Computing*, New York, NY, USA, 2008, pp. 6:II.1–6:1
13. G.E. Sakr, I.H. Elhajj, U.C. Wejinya, Multi level SVM for subject independent agitation detection, in *IEEE/ASME International Conference on Advanced Intelligent Mechatronics, 2009. AIM 2009*, 2009, pp. 538–543
14. N. Sebe, M.S. Lew, I. Cohen, A. Garg, T.S. Huang, Emotion recognition using a Cauchy Naive Bayes classifier, in *16th International Conference on Pattern Recognition, 2002. Proceedings*, 2002, vol. 1, pp. 17–20

Part III
Behavioural Measurement

Chapter 6
Linguistic Feature-Based Measures

Words, phrases and sentences are basic linguistic components that are used to exchange information between people. Recent investigations have revealed that people may demonstrate different linguistic patterns under levels of cognitive load. The adaptation to high cognitive load can be identified from longer sentences, more negative and less positive words, increased plural personal pronouns and decreased singular pronouns. These results have shown that cognitive load impacts the language capability of people, and high cognitive load may result in complex sentence structures that are difficult to understand. On the other hand, language complexity, especially in language rich environment, can be used as a potential measurement method to monitor cognitive load.

This chapter mainly involves two aspects of discussions regarding cognitive load measurement using linguistic features:

- Cognitive load measurement via words and non-word linguistics;
- Cognitive load measurement based on language complexities.

6.1 Linguistics

Linguistics deals with the study of language, and it involves three aspects: language form, language meaning and language in context [1]. Speaking practically, the field of linguistics analyzes human language as a system for sounds and meaning. Specifically, the use of particular words, sentences, syntax or grammatical structure, as well as the speaking style and pauses involved in speech are all included within the research scope of linguistics.

On the other hand, language is a coping mechanism in that it helps individuals lessen and manage both the causes and the effects of cognitive load [2]. Collecting the linguistic information is very convenient and can be conducted in real time with

F. Chen et al., *Robust Multimodal Cognitive Load Measurement*, Human–Computer Interaction Series, DOI 10.1007/978-3-319-31700-7_6

microphones, which makes a linguistic analytical method an ideal candidate for non-intrusive cognitive load measurement.

According to the complexity of sentence structure, linguistic features can be classified into three categories: non-word linguistics, words and sentence complexity. In the following sections they will be assessed respectively for their usefulness in cognitive load measurement.

6.2 Cognitive Load Measurement With Non-Word Linguistics

Non-word linguistics refers to the sections of speech for which there is no specific meaning, however they constitute important part of speech and help to keep the rhythm of speech. A typical example of non-word linguistics is pause, which is very common in everyday speech. Pause in speech is a mechanism that allows for more planning time, proper word selection and production of speech [3], [4].

Two different types of pause exist in speech:the silent pause and filled pause. A silent pause refers to the speechless or unvoiced pause segments, which usually occur between sentences, however in some cases it may occur within a sentence. The other type of pause is a filled pause. As its name suggests, a filled pause is not silent, although the voice still does not carry a specific meaning. Typical examples of filled pauses are *aah*, *umm*, *hmm*, etc. Sometimes it may be very difficult to differentiate between a silent pause and a filled pause, or the two types of pauses may occur together. A common method is to apply the Voice Activity Detection (VAD) method which is commonly adopted in speech signal processing, to discriminate the two types of pauses.

Pauses inherently originate from breathing activity and are often very brief. Therefore, to capture the pause as a feature, it is important to define a cut-off value, or threshold for its length. Although selected arbitrarily, this value usually ranges from 0.25 to 0.3 of a second [5].

Several categories of features can be extracted from pauses, including length of pauses, percentage of pausing time, frequency of pauses and response latency. They are addressed respectively as follows.

As shown in Fig. 6.1, T_p is the length of pause within the speech, and percentage of pausing time can be calculated as:

$$P_p = \frac{T_p}{T_s} \tag{6.1}$$

where T_s is the total speaking time.

Frequency of pauses can be calculated via the total number of pause occurrences divided by the total speaking time. Response latency is a task-dependent measure. Specifically, it is the time between people understanding the cognitive task and the onset of the time when people conduct the cognitive task. For example, in a reading

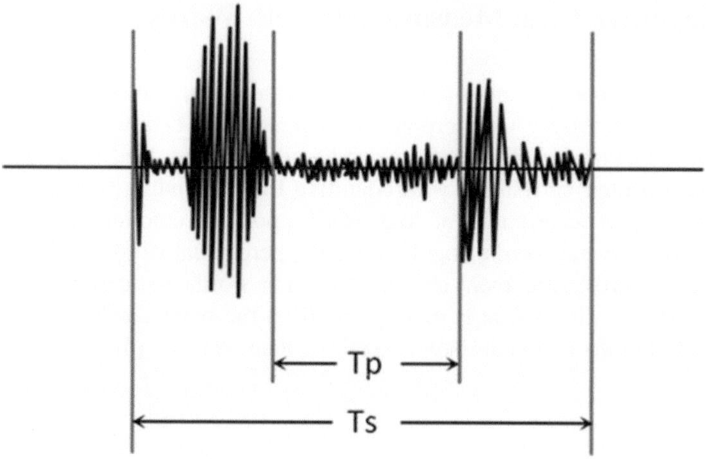

Fig. 6.1 Pause occurring in speech. T_p is the pause time, and T_s is the total speaking time. For the starting part of the speech, although no voice is produced, is not considered as pause as it is shorter than 0.25 s

comprehension task, the subjects are asked to read one paragraph of text, and answer some given questions. Every time the subject is asked a question, the time that elapses between the ending time of the question and the starting time of answering by the subject is considered the response latency.

According to the investigations of M. Asif Khawaja, it has been observed that the average pause lengths and percentages of time pausing increased with higher cognitive load, and the average pause frequencies and response latency showed the same trend as cognitive load was increased [6]. His further examination has shown that only silent pausing time increased significantly, while the filled pauses did not show a significant difference. His interpretation of this phenomena was based on the Baddeley's model [7], which involves the phonological loop that is responsible for language processing. When the phonological loop is overloaded, a subject is less likely to emit a filled pause than a silent pause. This may be because the execution of a filled pause itself may take up some aural/verbal resources for planning and execution in the subject's working memory, and the demand cannot be accommodated due to overload. Silent pauses, on the other hand, do not require extra processing power, and may be symptomatic of extra resources spent on handling internal cognitive processes such as response planning rather than response production. This may account for the way speech features are affected at the physical signal or surface level.

6.3 Cognitive Load Measurement with Words

Words are basic components of speech, and intuitively, it is expected that more words will be necessary to interpret and understand a complex problem than a simple one. In this section several word-based cognitive load measures will be introduced for their correlation with cognitive load, including word count, words per sentence, the occurrences of long words, positive emotion words, negative emotion words, swear words, cognitive words, perceptual words, inclusive words, achievement words, agreement and disagreement words and certainty and uncertainty words. The discussion is mainly based on the work conducted by M. Asif Khawaja during his linguistic-based cognitive load research [6].

6.3.1 Word Count and Words per Sentence

It has been observed that under high cognitive load conditions, people tend to speak more with each other, and hence the total words spoken can be an indicator of cognitive load. Furthermore, longer sentences also occur together with more communications under high cognitive load conditions.

6.3.2 Long Words

Based on discussions above, as cognitive load increases, people have used more complex languages with increased sentence length. On the other hand, size of the spoken words, in terms of the number of letters involved, can also be a parameter depicting language complexity. Practically speaking, a 'long word' does not have a strict definition, and therefore we consider words involving at least six letters to be long words. As expected, more long words can be identified spoken by people under high cognitive load conditions compared with low cognitive load conditions.

6.3.3 Positive and Negative Emotion Words

Positive and negative emotion words are frequently used in everyday speech. For example, *good*, *happy*, *interesting*, *fine*, *perfect* etc. represent a person's positive feeling of amusement and happiness, and thus these words are considered as positive emotion words. In comparison, words like *angry*, *hate*, *blame*, *afraid*, *worried* etc. are utilized when people have a negative feeling such as fear, sadness, or frustration. These words are categorized as negative emotion words.

The latest investigation based on a bushfire study [8] has shown that the usage of positive emotion words will drop as cognitive load increases, while negative emotion words will be more frequently used under high cognitive load conditions. This finding is interesting as it reveals the possible link between cognitive load and emotion, as reflected from the choice of words in the speech.

6.3.4 Swear Words

Similar with negative emotion words, it has been observed that under extreme frustrating situations which usually accompany high cognitive load, people tend to use more swear words or expletives. However there was not a significant difference between the low and high cognitive load conditions as demonstrated by the negative emotion words in the same study [6].

6.3.5 Cognitive Words

Cognitive words represent the human cognitive process, and are mostly verbs such as *consider*, *know*, *understand*, *identify* etc. Sometimes some adjectives can be used, such as *convinced*, *sure*, *confident* etc. Observations have shown that more cognitive loads will be used when subjects are conducting tasks with high cognitive load, reflecting their meditation process.

6.3.6 Perceptual Words

Perceptual words represent a person's perception process. For example, *feel*, *touch*, *observe*, *hear*, *view* etc. Perceptual words are quite similar with cognitive words, and thus more usage of perceptual words are expected under high cognitive load conditions, because people need to concentrate more on the high-load task compared with the low-load tasks, in order to maintain the task performance.

6.3.7 Inclusive Words

Words like *along*, *with*, *both*, *all*, *including*, *everyone* are typical examples of inclusive words, which designates the range of involvement. More use of inclusive words has been reported under high cognitive load, showing people's increasing linguistic richness in an effort to communicate effectively between themselves and other collaborative partners.

6.3.8 Achievement Words

Achievement words, for example, *win*, *victory*, *reward*, *hero* etc., represent people's feeling of achievement. Although there was a decreasing usage of achievement words reported in a recent study by M. Asif Khawaja [6], the statistical examination in the same study didn't reveal a significant difference between high and low cognitive load conditions.

6.3.9 Agreement and Disagreement Words

Words such as *OK*, *agree*, *correct*, *exactly* etc. are considered agreement words, and words like *no*, *not*, *never* are examples of disagreement words. It is proposed that under a group collaboration environment, more agreement words occur when the cognitive load is low, but more disagreement words are spoken when the cognitive load is high [6]. This means that people show their assent to each other much more readily under low cognitive load tasks than under high cognitive load tasks, where they would more likely show their opposition on the decisions made by other team members.

6.3.10 Certainty and Uncertainty Words

Certainty words and uncertainty words represent a person's confidence and understanding of the actions taken or the decision made. Typical certainty words include *sure*, *right*, *absolutely*, *definitely* etc., and uncertainty words may include *doubt*, *guess*, *might*, *perhaps*, *possibly* etc. It is likely that people use more certainty words instead of uncertainty words under low cognitive load conditions, while use more uncertainty words under high cognitive load conditions due to decreased confidence about what and how they are performing.

6.3.11 Summary of Measurements

The summary is based on the examination of M. Asif Khawaja on two datasets, one from simulated laboratory study (controlled study) and the other from real-life study, both of which are in the bushfire emergency service context, as tabulated in Table 6.1 [6, 8, 9].

It is well established that people's selection of language elements and linguistic features, as well as speech production mechanism vary from one situation to another, depending on the conditions and the circumstances of the situation

Table 6.1 Linguistic features for cognitive load examination. 'Yes' means significant difference between high and low cognitive conditions has been identified for a given feature

Linguistic features	Behavior similar in both studies?	Significant in controlled study?	Significant in real-life study
Words per Sentence	Yes	Yes	Yes
Positive Emotions	Yes	Yes	Yes
Cognitive Words	Yes	Yes	Yes
Perceptual Words	Yes	Yes	Yes
Agreement Words	Yes	Yes	Yes
Disagreement Words	Yes	Yes	Yes
Word Count	Yes	Yes	
Long Words	Yes	Yes	
Inclusive Words	Yes	Yes	
Negative Emotions	Yes		Yes
Swear Words	Yes		
Achievement Words	Yes		
Certainty Words	Yes		
Uncertainty Words	Yes		

[2, 10]. Although not every feature mentioned in the above discussion demonstrates significant effect under different cognitive load conditions as shown in Table 6.1, most of the features have been identified to be effective indicators of cognitive load levels.

However, it should be noted that the above word-based features, although defined independently, there do exist some overlaps between them. For example, the word *definitely* can be both a certainty word and an agreement word, and thus a refined word categorization system may be helpful in cognitive load examination. Furthermore, the cognitive words often constitute parts of pet phrases, and some people would like to start a speech with these words even if there is no cognitive process involved. These issues should also be considered in future linguistic-based cognitive load examinations.

6.4 Cognitive Load Measurement Based on Personal Pronouns

In most human languages, pronouns exist as substitutes for a noun. A personal pronoun is a word that substitutes for the designated people. Due to the special properties and frequent usage of personal pronouns, they are discussed separately to the words in the previous section. Specifically, three categories of personal pronoun will be addressed in this section, including:

- *Singular personal pronouns*: for example, I, she, he, his, etc.
- *Plural personal pronouns*: for example, we, they, their, etc.
- *Hybrid personal pronouns*: for example, you, yours.

The three categories of personal pronouns were further divided into five pronoun sub-types as shown in table Table 6.2, where each pronoun type encompasses all possible personal pronouns relevant to that type. The pronoun type *You* can be used as both singular and plural pronouns referring to either a single person or a group of people.

However, according to M. Asif Khawaja's examination, overall, no significant main effects have been found between the total pronouns used and the cognitive load conditions. One interesting observation was that the use of singular pronouns decreases and the use of plural pronouns increases under high cognitive load tasks, although the total usage of pronouns remained stable across different cognitive load levels.

Based on the discussion above, participants, especially working in collaborative and team environments, consistently use singular pronouns and plural pronouns differently under different cognitive load conditions. This observation suggests that when dealing with low cognitive load tasks, team members feel more confident about the task and prefer to perform tasks individually. On the other hand, when dealing with complex and high cognitive tasks, they feel less confident and thus more interactions with other team members become necessary, to resolve the problem efficiently.

Although the pronoun 'you' consistently decreased from low to high cognitive load conditions, it is difficult to interpret the corresponding behavior accordingly, due to the fact that it could be either singular or plural.

Table 6.2 Types of personal pronouns

Pronoun category	Sub-type	Examples
Singular Pronouns	I	I, me, mine my
	She/He	She, his, her, him
Plural Pronouns	We	we, us, our, ours
	They	they, their, them
'Hybrid' Pronouns	You	you, yours, your

6.5 Language Complexity as Indices of Cognitive Load

Long sentences have more chances to contain more clauses, tense and semantic variations, and therefore they are capable of communicating more information and varying ideas. As a consequence, the longer the sentence is, the more complex the language and the more difficult to comprehend it. Together, all these factors contribute to the complexity of a text or transcript.

The complexity of language mainly refers to criteria related to the construction of a sentence via the selection of words and the construction of sentences. Specifically, it involves two factors, ie semantic difficulty and syntactic complexity [11].

Semantic difficulty describes the use of words, their structure and length (both in syllables as well as letters or characters), while syntactic complexity primarily observes the length of sentence, which is considered the best indicator of language complexity.

Specifically, both categories of complexity measures will be discussed in this section, including lexical density, complex word ratio for semantic difficulty, Gunning Fog Index, Flesch-Kincaid Grade, and the SMOG Grade for syntactic complexity of sentences. Most of these measures have been developed by researchers to evaluate the readability of written text and to conduct statistical surveys on readers of a text [12].

6.5.1 *Lexical Density*

Lexical density is the estimated measure of content per functional and lexical units or lexemes in a complete text [13]. In brief, it is a measure of the ratio of unique or different words to the total number of words and denotes the vocabulary richness of the subject. Practically, Lexical Density (*LD*) can be calculated as:

$$LD = \frac{n_{dw}}{N} \times 100 \tag{6.2}$$

where n_{dw} is the total number of different words, and N is the total number of words.

6.5.2 *Complex Word Ratio*

Complex words are the words with three or more syllables [14]. They are also known as polysyllables. Each syllable in a word refers to a sound that can be spoken without interruption, and is usually a vowel which can have consonants before or after it. For example, the word ‘density’ has three syllables.

Complex Word Ratio (*CWR*) is the measure of the ratio of complex words to the total number of words and can be calculated as follows:

$$CWR = \frac{n_{cw}}{N} \times 100 \quad (6.3)$$

where n_{cw} is the total number of complex words or polysyllables, and N is the total number of words.

6.5.3 Gunning Fog Index

The Gunning Fog Index (*GFI*) calculates the syntactic complexity of the language using sentence lengths and complex words and implies that short sentences in plain English achieve a better score than long sentences in complicated language [12, 15, 16]. Specifically, it can be calculated as:

$$GFI = 0.4 \times \left(ASL + \frac{n_{cw}}{N} \times 100\right) \quad (6.4)$$

where *ASL* is the Average Sentence Length.

6.5.4 Flesch-Kincaid Grade

The Flesch-Kincaid Grade (*FKG*) describes language complexity using average sentence lengths and average syllables per word [17]. It estimates the number of years of education required to understand the written or transcribed text, and can be calculated as follows:

$$FKG = 0.39 \times ASL + 11.8 \times ASW - 15.59 \quad (6.5)$$

where *ASW* indicates the Average number of Syllables per Word.

6.5.5 SMOG Grade

The SMOG Grade (Simple Measure Of Gobbledygook) also estimates the number of education years needed to fully comprehend the text [18]. It uses number of sentences and complex words to calculate it. The emphasis on full comprehension distinguishes this measurement from other complexity measurements. The SMOG Grade can be calculated as follows:

$$SMOG = \sqrt{\frac{n_{cw}}{n_s} \times 30} + 3 \quad (6.6)$$

where n_s refers to the total number of sentences.

Table 6.3 Complexity measurements and corresponding linguistic factors involved in calculation

Complexity Measure \ Linguistic Factor	Sentence Length	Number of Words	Punctuations	Syllables	Complex Words	Full comprehension
Lexical Density		✓				
Complex Word Ratio		✓			✓	
Gunning Fog Index	✓	✓	✓	✓	✓	
Flesch-Kincaid Grade	✓	✓	✓	✓		
SMOG Grade			✓	✓	✓	✓

6.5.6 *Summary of Language Measurements*

The following table summarizes the factors involved in the different linguistic measurements [6] (Table 6.3).

A comparative study conducted by M. Asif Khawaja has shown that the linguistic measures are capable of discriminating the different cognitive load conditions. Specifically, it was observed that subjects' lexical density, which represents a person's vocabulary richness, significantly decreased when the cognitive load was increased. Also, the subjects' overall conversational language was found to be increasingly complex, ie difficult to understand under high cognitive load conditions. The high values of GFI, Flesch-Kincaid and SMOG seen under high cognitive load conditions demonstrated that the subjects' language became more complex and consisted of longer sentences, longer words, more clauses, punctuations and varying ideas compared with those of low cognitive load conditions. It was also noted that subjects used an increased number of complex or polysyllable words under high cognitive load conditions than low cognitive load conditions.

6.6 Summary

This chapter discussed methods for cognitive load measurement via linguistic features. It can be found that many linguistic features, which were originally designed to examine language complexity for learning analytics or comprehension, can be successfully applied to the cognitive load research. The linguistic assessment of cognitive load through analysis of user's speech is attractive because it offers the potential to achieve adaptive system behavior. If users experiencing high cognitive load can be identified by the system, they can be catered for with extra support if necessary, or through adaptation of the organizational or system behavior to decrease their overall experienced cognitive load to more manageable levels. Furthermore, the linguistic approach to measure cognitive load can be used as a post-hoc analysis technique for user interface evaluation and interaction design improvement, as long as the scripts of speech is available.

References

1. M.A.K. Halliday, J.J. Webster, M.A.K. Halliday, *On Language and Linguistics* (Continuum, London, 2003)
2. J.B. Sexton, R.L. Helmreich, Analyzing cockpit communication: The links between language, performance, error, and workload. J. Hum. Perform. Extrem. Environ. **5**(1), 63–68 (2000)
3. J. Schilperoord, On the cognitive status of pauses in discourse production, in *Contemporary Tools and Techniques for Studying Writing* (Kluwer Academic Publishers, London, 2001)
4. K.M. Rhee, E. Kim, A statistical analysis of text for inferring authenticity, in *Proc. 53rd Session of International Statistical Institute*, 2001
5. J. Schilperoord, *It's About Time: Temporal Aspects of Cognitive Processes in Text Production* (Rodopi, Amsterdam/Atlanta, 1996)
6. M.A. Khawaja, *Cognitive Load Measurement using Speech and Linguistic Features* (University of New South Wales, Sydney, 2010)
7. A.D. Baddeley, Working memory. Science **255**, 556–559 (1992)
8. M.A. Khawaja, F. Chen, N. Marcus, Measuring cognitive load using linguistic features: Implications for usability evaluation and adaptive interaction design. Int. J. Hum. Comput. Interact. **30**(5), 343–368 (2014)
9. M.A. Khawaja, F. Chen, N. Marcus, Using language complexity to measure cognitive load for adaptive interaction design, in *Proceedings of International Conference on Intelligent User Interfaces (IUI 2010)*, Hong Kong, China, 2010, pp. 333–336
10. H.W. Dechert, M. Raupach, *Towards a Cross-Linguistic Assessment of Speech Production* (Lang, Frankfurt, 1980)
11. C. Lennon, H. Burdick, *The Lexile Freamework as an Approach for Reading Measurement and Success* (MetaMetrics, Inc, Durham, 2004)
12. R.P. Reck, R.A. Reck, Generating and rendering readability scores for Project Gutenberg texts, in *Proceedings of the Corpus Linguistics Conference*, Birmingham, UK, 2007
13. J. Ure, Lexical density and register differentiation, in *Applications of Linguistics*, ed. by G. Perren, T. Trim (Cambridge University Press, London, 1971), pp. 443–452
14. S. Chalker, E. Weiner, *The Oxford Dictionary of English Grammar* (Oxford University Press, Oxford, 1998)
15. Advanced Text Analyser. UsingEnglish, http:
16. R. Gunning, *The Technique of Clear Writing* (McGraw-Hill, New York, 1952)
17. R. Flesch, A new readability yardstick. J. Appl. Psychol. **32**(3), 221–233 (1948)
18. H.G. Mclaughlin, SMOG grading – a new readability formula. J. Read. **12**(8), 639–646 (1969)

Chapter 7
Speech Signal Based Measures

Speech is an ideal modality for cognitive load measurement, due to its universal availability, relatively low cost to acquire, low processing requirement for devices and can be largely non-intrusive. A speech-based cognitive load measurement system might potentially be able to continuously monitor the cognitive load of a person without interfering with the task being conducted. One example of such a system that monitors cognitive load in a non-intrusive way, designed specifically for call centers, is a product called BrainGauge.

Many speech parameters have been found to be affected by cognitive load, including prosody, spectrum and vocal tract related parameters. For example, Berthold's investigation has shown that features such as the number of sentence fragments (eg whole sentences versus incomplete sentences), self-interruptions and occurrence of self-repairs when speaking, as well as articulation rate were reflective of cognitive load variations [1]. Similar research conducted by Mueller also confirmed that speaking disfluencies, articulation rate, utterance content quality, complexity of sentence construction as well as silence and filled pauses in an utterance could be used to detect cognitive load via speech [2].

Accordingly, various classification methods have been applied to cognitive load examination. Typical methods include Gaussian Mixture Models (GMM), linear kernel Support Vector Machine (SVM) and hybrid SVM-GMMs which take the output of a GMM as the input for a SVM. Channel and speaker normalization is also widely adopted, together with different delta techniques. Furthermore, a Universal Background Model (UBM) is often necessary to reduce the impact of insufficient data available.

This chapter will review the recent investigations into cognitive load measurement via speech, which mainly involve:

- Experiments utilized to induce different levels of cognitive load;
- The various speech features extracted for cognitive load investigation;
- The comparison of different classification methods.

F. Chen et al., *Robust Multimodal Cognitive Load Measurement*, Human–Computer Interaction Series, DOI 10.1007/978-3-319-31700-7_7

Apart from the discussions of cognitive load measurement, cognitive load-aware system design issues are also discussed with an emphasis on real-time cognitive load measurement and its implication on system usability.

7.1 Basics of Speech

Speech is a natural form of communication for human beings. Although the main objective of speech is to convey linguistic information, this is not the only information conveyed by speech. Other information including speaker identity and mental state related information such as cognitive load is also conveyed in speech [3]. Speech is an acoustic signal, generated by the airflow from the lungs (considered to be the voice source) which then passes through the pharynx and then the oral and nasal cavities, collectively known as the vocal tract filter, as shown in Fig. 7.1. Normally the generation of voice can be modelled in a simplified version following the pathway of air, as depicted in Fig. 7.1b. The parameters of the voice source and the vocal tract filter vary according to the content of the utterance to be pronounced as well as the mental state of the speaker.

7.2 Cognitive Load Experiments

A general requirement for a cognitive load experiment is that it should be capable of inducing varied cognitive load in a controlled way. Specifically, the experiment should involve tasks of varied difficulty, which correspond to different levels of cognitive load. The task difficulty should be the main cause of cognitive load variations, which means the implications of other factors should be excluded to the extent that their effect on cognitive load can be omitted. Additionally, speech should be the major modality of interaction, which implies that the experiment participant is expected to conduct the task via speech instead of other methods such as mouse or gesture, which may overshadow the cognitive load effect during the conduction of the experiment.

Several typical experiments will be introduced here, including a reading and comprehension task, a Stroop Test and a reading span task. Each of these methods will be addressed in detail.

7.2.1 *Reading Comprehension Experiment*

The general task in a reading comprehension experiment requires a participant to complete two sub-tasks in sequence. The first sub-task is to read a short passage, the difficulty level of which is typically assessed with the Lexile Framework for

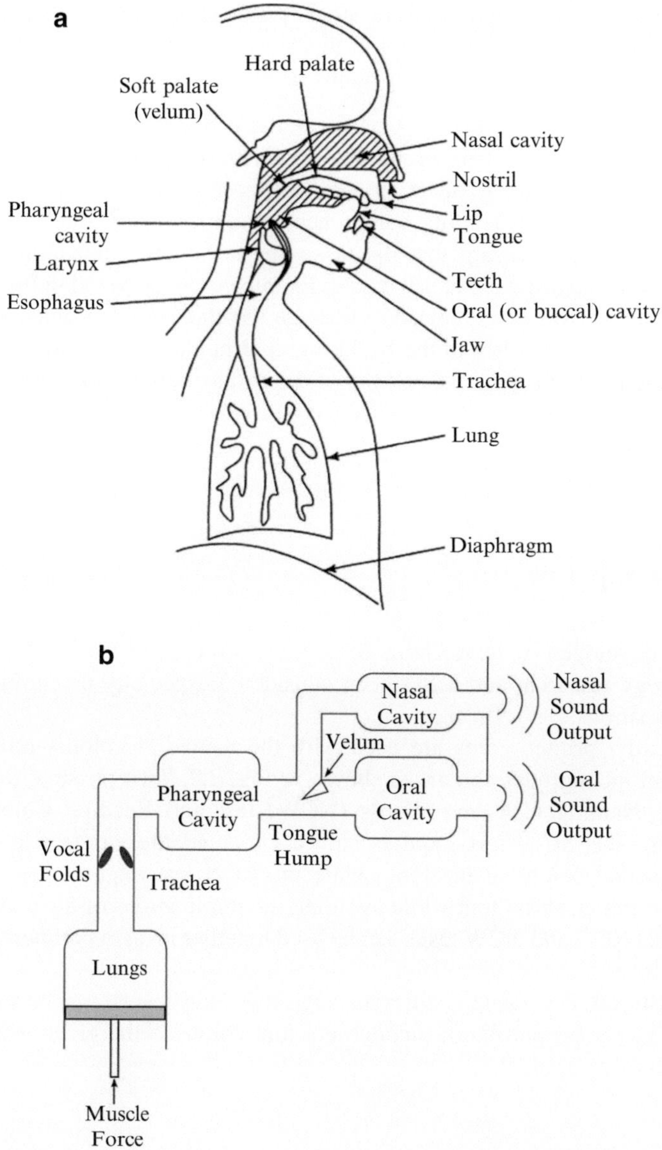

Fig. 7.1 (**a**) Organs related to the generation of speech [4] and (**b**) a simplified model of voice generation ([4], copyright © 2000 Wiley, reprinted by permission of Publisher of Wiley)

Reading [5], which evaluates the semantic difficulty and syntactic complexity of the text. For a beginner's level text the score might be 200 L (Lexiles), while for an advanced level of reading, a score over 1700 L could be applicable. The other sub-task is then conducted via a Q&A session, and three typical open-end questions can be:

- Please give a short summary of the story in at least five whole sentences.
- What is the most interesting point in this story?
- Please describe at least two other points heighted in this story.

The participants are expected to answer these questions via speaking. Even though there is not correct or wrong answer for these questions, this is expected to be an effective means of collecting enough speech data to assess cognitive load.

In some examples of reading comprehension experiments, dual task schemes have been applied to ensure that the tasks are sufficiently demanding and thus sufficiently high cognitive load is induced. Examples can be found in [6] and [7], in which the subject was given a headset through which a series of random two digit numbers were played softly in the background at random intervals, while participants were required to count the number of digits that were heard – concurrently to the reading task. A dual task approach may also be applied during the question answer session.

7.2.2 *Stroop Test*

Originally developed by John Ridley Stroop, the Stroop Test has been widely used in psychology research, and it has been considered especially useful in cognitive load examinations.

Specifically, printed cards are used with the names of colours printed using letters of an incongruent colour, as shown in Fig. 7.2. Two possible tests can be applied, ie Naming Coloured Words (NCW) test and Reading Colour Names (RCN) test. The NCW test requires subjects to read the words without caring about the actual colour of the font, while the RCN test requires the subjects to identify the colour of the font while avoiding the implications of the text. It should be noted that RCN and NCW tasks can be used together in an experiment, as shown in Table 7.1.

In the Stroop Test design, different cognitive load levels can be induced via manipulating the proportion of incongruent font colour in the given words, as the example show in Table 7.1.

7.2.3 *Reading Span Experiment*

The reading span experiment is commonly used to measure the working memory capacity of a subject [8–10]; which is predicated on the maximum number of items that can be held inworking memory. Due to the close relationship between working memory and cognitive load, a more challenging task that requires more working memory results in higher cognitive load. Another advantage of the experiment

Yellow Green Red Blue Red Purple

Fig. 7.2 Exemplary cues for a Stroop Text. The names of the colours are different from the font of the text. The subject is asked to speak the colour of the font out loud, ie the 'Naming Coloured Words' test, or read the text irrespective of the font colour, ie the 'Reading Colour Names' test

Table 7.1 Exemplary Stroop Test tasks to induce three levels of cognitive load

Level	Type	Test design
Low	RCN	All words written in the same colour, or
		All words written in congruent font colour
Medium	NCW	50 % words written in a congruent colour
High	NCW	All words written in an incongruent colour

employing the number of items stored in working memory is that it provides a convenient means of quantification.

Usually each task in the reading span experiment is implemented as dual task format, with the reading task as the primary task while a secondary task such as memorizing a given letter is required. The reading task may require the subject to make judgement on whether the sentence is logical or not, for example, a given sentence like "I like to walk in the sky." can be considered illogical, while "I like to walk along the silent path." might be a logical sentence. The main independent factor in the reading span experiment is the level of content required to read, which can be quantified in terms of the number of sentences, or the total number of letters involved.

7.2.4 Time Constraint

Sometimes to induce additional levels of cognitive load, a time constraint is applied in the experiment, especially to the high cognitive load tasks. Asking the participant to finish a particular task within a limited time is considered likely to result in stress in the participant, since stress is commonly linked to high levels of cognitive load [11]. Time constraints can be a convenient means, applicable to any task, to further increase the difficulty and induce extreme cognitive load.

However, several issues should be considered prior to implementing timing pressure within an experiment. Firstly, compared with the factors such as the number of given words in a reading comprehension experiment or a reading span experiment, time is an independent variable from another dimension, which means that time cannot be quantified in the same way as other parameters of the experiment. Secondly, a time constraint applied to different subjects may not result in the same stress effect due to individual differences, and thus may not have the same implications on induced cognitive load. For example, with a time constraint of 2 s in a recall task, one experienced subject may be able to conduct the task within 0.5 s, while another novel subject may need more than 3 s to cope with the task. In such a case the time constraint is not effective for the more capable subject; however it is likely to have an effect on the less capable subject. Furthermore, time constraints may cause other problems in terms of experimental protocol, and the experiment designer will have to decide what will happen if the subject cannot complete the task within the given time slot: will next task pop up unexpectedly, or will the current task remain visible? In the former case the participant may be obliged to skip some tasks which may ultimate result in incomplete data, while in the latter case the time constraint actually loses its intended effect. Finally, especially for speech-based cognitive load evaluation, time constraints may cause frustration, which, together with stress, may prompt the subjects to change the style of speaking, but not necessarily resulting from cognitive load. Normally a time constraint may result in increased speaking speed, however whether the speed change can be an indicator of cognitive load under such experiment conditions remains arguable.

7.2.5 Experiment Validation

Varied cognitive load is expected to be induced in the aforementioned experiments, and in consequence a validation process is usually conducted to ensure high cognitive load is experienced by subjects in the difficult tasks, while low cognitive load occurs for the easy tasks.

The validation process is usually conducted via two methods: subjective ratings and performance on the tasks. Subjective ratings are one of the most common methods to examine cognitive load as addressed in Chap. 2. Performance-based measurement is dependent on the experimental setting, and different experiments are likely to have different performance examination criteria, for example, the accuracy of summarization in the reading comprehension experiment, or the error rate of the Stroop Test are typical performance measurements of cognitive load.

7.3 Speech Features and Cognitive Load

As discussed before in Sect. 7.1, human speech production involves the signal source, ie the air passing through the lung, and the filter, which is dependent on the overall structure of the vocal tract. As the impact of cognitive load on the human speech production system can occur at the source, the filter or both, it is feasible to extract source-based or filter-based feature or a combination of the two categories of features for cognitive load examination.

7.3.1 Source-Based Features

- Pitch
  Pitch is a feature that characterizes the vibration rate of the vocal folds. Many methods can be used to calculate pitch, and the most population ones include auto-correlation method, and the Robust Algorithm for Pitch Tracking (RAPT) proposed by Talkin [12].
- Intensity
  Intensity refers to the amplitude of the vocal fold vibration which in turn depends on the pressure of the subglottic airstream. Practically, the loudness of speech is determined by the sound pressure level of the sound wave at the listener's eardrum. The pressure level depends on the intensity of the speech at the mouth of the speaker and the distance between speaker and listener. In speech analysis, the loudness of recorded speech is determined via the sound pressure level at the microphone. The intensity is dependent on this loudness and the transfer function of the microphone.
- SourceMel Frequency Cepstral Coefficients (SMFCC)
  This feature represents the spectral envelope of the glottal waveform in a compact format. A series of triangle filters equally spaced in the Mel frequency scale are applied on the glottal waveform to extract this feature (Fig. 7.3).

Although both intensity and pitch are source-based features, according to [13] they capture different aspects of the voice source. The cognitive load information contained in these two features can therefore complement each other. On the other hand, studies conducted on the same data set by Phu have shown that the SMFCC features are capable of reaching good cognitive load level classification performance as well, which indicates that voice source related features are suitable for cognitive load classification.

7.3.2 Filter-Based Features

- Formant frequencies
  When the excitation signal passes through the vocal tract it is shaped by the resonance characteristics of the vocal tract, which results in a number of peaks in the magnitude spectrum of the speech signal. The resulting peaks, referred as speech formants, are located at the resonant frequencies of the filter modeling the vocal tract.
- Filter Mel Frequency Cepstral Coefficients (FMFCC)
  As a compact representation of the spectral envelope of the vocal tract filter, the FMFCC feature can be via estimating the effect of the voice source on the speech signal. Similar with SMFCC, a series of Mel filters are applied to the spectral envelope, and FMFCCs are the coefficients of the discrete cosine transform of the logarithm of the filter output.

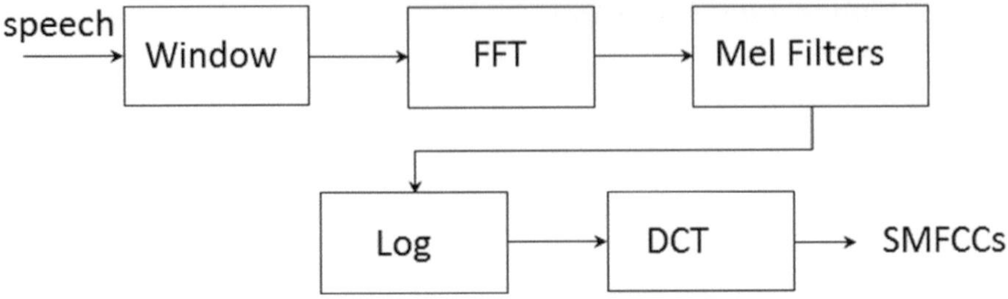

Fig. 7.3 Procedure to extract SMFCC features. The speech signal is first segmented using a sliding window, and Fast Fourier Transform (*FFT*) is applied to the speech signal within each window. The result is filtered with Mel Filters, and the logarithm of the output is calculated and finally a Discrete Cosine Transform (*DCT*) is utilized to calculate the final SMFCC features

- Electroglottograph features
  A typical electroglottograph (EGG) device consists of a pair of electrodes that are placed on both sides of the neck close to the vocal folds. As the vocal folds open and close, the change in impedance resulting from the surrounding tissues can be captured by the EGG device. Using the EGG signal to detect cognitive load is mainly based on the first-order derivative of it.
- Glottal flow features
  As shown in Fig. 7.4, a typical glottal flow involves four phases. It is the volume velocity flowing through the glottis, or the excitation source of voiced speech in the source-filter model. Specifically, during the closed phase the glottal flow is zero as the vocal folds are closed. The opening phase indicates that the glottal flow increases due to the open vocal folds, and maximizes when the vocal folds are open to its maximum level. The closing phase indicates the glottal flow decreases due to the closing of vocal folds. The return phase follows the closing phase until the glottal flow reaches zero.

According to Phu's investigation [13], the filter-based features overall performed better than source-based features, which indicates that cognitive load has a higher implication on the vocal tract than the source of the speech. Furthermore, his examination also suggested that combining speech features from the source and the vocal tract parameters is capable of producing better cognitive load classification results, which implies that different speech features characterize different aspects of the speech production system and thus contribute to cognitive load information in a complementary way. Tet's investigation also showed that the formant information is useful for cognitive load classification. Furthermore, Tet has identified that as cognitive load increases, vocal folds open and close faster [14]. Additionally, vocal folds stay open for shorter periods of the glottal cycle when cognitive load is increased.

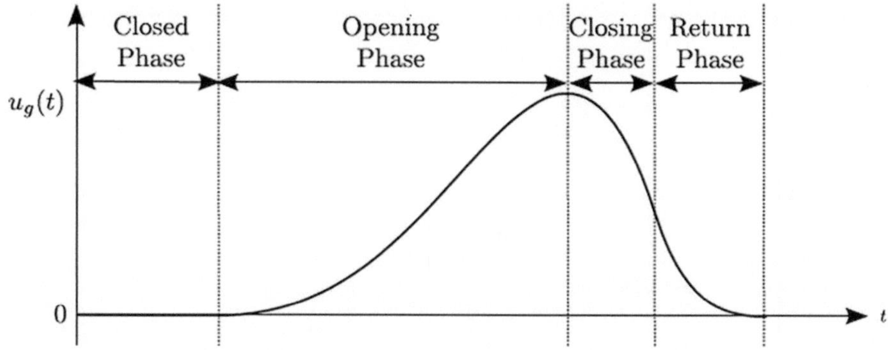

Fig. 7.4 A typical glottal flow waveform

7.4 A Comparison of Features for Cognitive Load Classification

In view of the many features of speech proposed as being able to classify cognitive load levels, it may be considered prudent to conduct a quantitative comparison between these different features on the same dataset in order to examine their respective performance, which is conveniently supplied by Tet's examination [14].

Specifically, Stroop Tests involving three cognitive load levels as discussed in Sect. 7.2.2 were carried out. Each load level consisted of three trials, and in each trial participants were asked to name 20 colours. Prior to the experiment a practice trial consisting of 10 colour words was applied. The sequence of colour names/words was randomly presented. Also, after each trial, the participant rated the experienced cognitive load on a scale of 1 (lowest cognitive load) to 9 (highest cognitive load), designed as a validation measure. A total of 216 utterances were collected in the dataset, which involved three cognitive load levels with 72 utterances in each. On average the duration of each utterance was 16.6 s.

7.4.1 Pitch and Intensity Features

As shown in Fig. 7.5, pitch increases according to cognitive load level, which is supported by the results reported in [15–18].

When the pitch was extracted from the speech of the Stroop Test, it was found that pitch alone was not quite ideal for cognitive load classification and the performance was close to chance level, however pitch and its shifted delta feature was found to be a better choice compared with pitch alone [13], as shown in Table 7.2.

Like pitch, intensity is a single dimension feature and its performance in discriminating three cognitive load levels was tabulated in Table 7.3. Generally

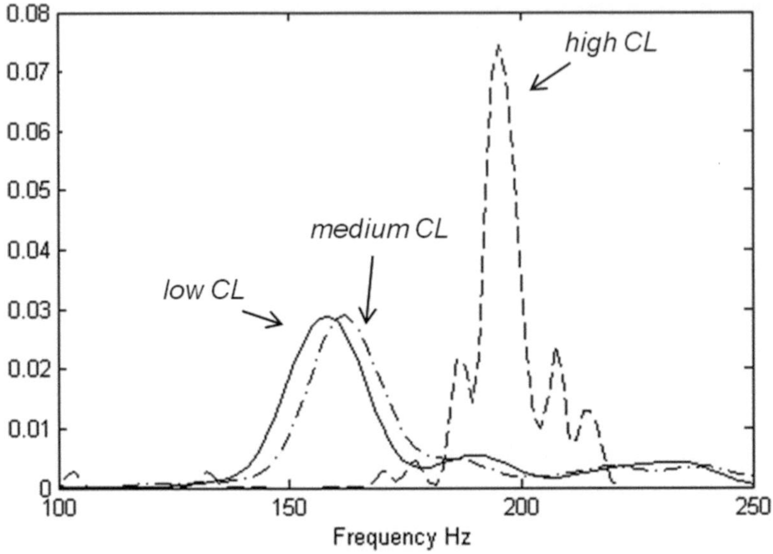

Fig. 7.5 Pitch distribution of the word ‘gray’ under different cognitive load conditions

Table 7.2 Classification accuracies of the system using pitch

Corpus	Accuracy (%) Pitch	Pitch and its shifted delta feature
Stroop test	32.8	52.2
Reading and comprehension	33.3	37.0

Table 7.3 Classification accuracies of the system using intensity

Corpus	Accuracy (%) Intensity	Intensity and its shifted delta feature
Stroop test	32.8	56.9
Reading and comprehension	34.1	41.5

the intensity feature performed similarly to pitch; however it should be noted that although both pitch and intensity are both source-based features, they capture different aspects of the voice source and thus the incorporation of both features have been shown to improve the overall performance of the cognitive load classification system [14].

7.4.2 EGG Features

For a given speech shown in Fig. 7.6, the EGG and DEGG features can be calculated as follows:

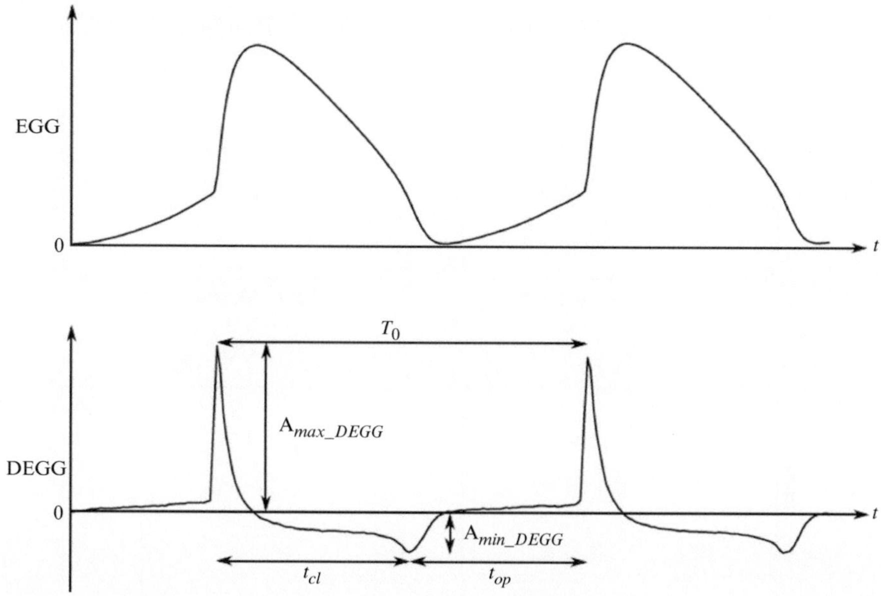

Fig. 7.6 Example EGG and DEGG signals, with time and amplitude instants used in calculating the EGG parameters [14]

Maximum positive peak of DEGG (A_{max_DEGG}): A_{max_DEGG} occurs when the rate of closing of the vocal folds is at a maximum level. A large value of A_{max_DEGG} indicates that the vocal folds are closing at a high speed.

Maximum negative peak of DEGG (A_{min_DEGG}): Similar to A_{max_DEGG}, A_{min_DEGG} occurs when the rate of opening of the vocal folds is at a maximum. The more negative A_{min_DEGG} is, the higher speed the vocal folds are opening.

EGG-based open quotient (OQ_{EGG}) and close quotient (CQ_{EGG}): The EGG-based features relates to the duration of the vocal fold opening at a proportion of the pitch period. The open quotient can be calculated as:

$$OQ_{EGG} = \frac{t_{op}}{T_0}$$

where t_{op} is the time interval between the glottal opening instant and the next glottal closure instant, and can be calculated as the time interval between the maximum negative DEGG peak and the subsequent maximum positive DEGG peak. A lower OQ_{EGG} value indicates that the vocal folds stay open for a shorter proportion of the pitch period.

The EGG features as described above have been shown to be reflective of cognitive load variations, and Fig. 7.7 illustrates the relationship between the

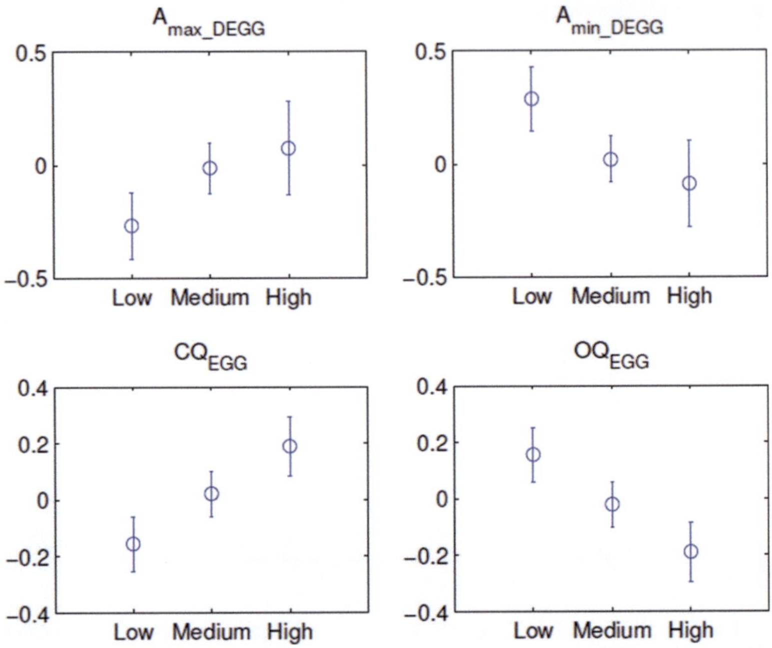

Fig. 7.7 Means and 95 % confidence intervals of EGG features under different cognitive load levels [14]

features and three cognitive load levels for the given speech data from the Stroop Test.

Using the features to classify three cognitive load levels, an average accuracy above 50 % was achieved, as shown in Table 7.4.

7.4.3 Glottal Flow Features

Glottal flow features were examined as well on the Stroop Test data. Specifically, according to Tet [14], nine features were examined as shown in Fig. 7.8. Specifically, the features include:

Primary open quotient:

$$OQ_1 = \frac{t_c - t_{o1}}{T}$$

Secondary open quotient:

Table 7.4 Cognitive load classification accuracies based on EGG features

Features	Classification accuracy (%)			
	Low load	Medium load	High load	Average
A_{max_DEGG}	68.1	47.2	37.5	50.9
A_{min_DEGG}	65.3	51.4	45.8	54.2
OQ_{EGG}	58.3	27.8	61.1	49.1

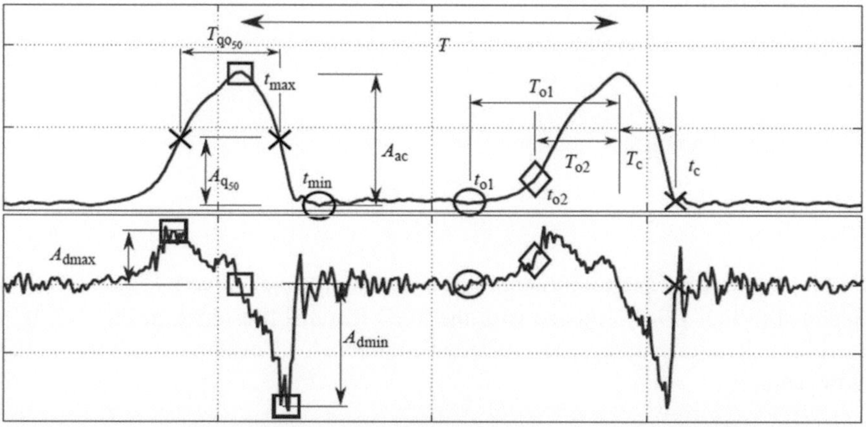

Fig. 7.8 Example glottal flow (*upper pane*) and glottal flow derivative (*lower pane*) waveforms, with time and amplitude instances used in calculating the time domain glottal flow parameters ([19], copyright © 2008 Taylor & Francis, reprinted by permission of Publisher of Taylor & Francis Ltd)

$$OQ_2 = \frac{t_c - t_{o2}}{T}$$

Amplitude-based open quotient:

$$OQ_a = \frac{A_{ac}}{T}\left(\frac{\pi}{2A_{dmax}} + \frac{1}{A_{dmin}}\right)$$

Quasi-open quotient:

$$QOQ = \frac{T_{QO50}}{T}$$

Primary speed quotient:

$$SQ_1 = \frac{t_{max} - t_{o1}}{t_c - t_{max}}$$

Secondary speed quotient:

$$SQ_2 = \frac{t_{max} - t_{o2}}{t_c - t_{max}}$$

Closing quotient:

$$ClQ = \frac{t_c - t_{max}}{T}$$

Amplitude quotient:

$$AQ = \frac{A_{ac}}{A_{dmin}}$$

Normalized amplitude quotient:

$$NAQ = \frac{AQ}{T}$$

The relationship between the glottal flow features and cognitive load levels is depicted in Fig. 7.9. Compared with the EGG features illustrated in Fig. 7.7, it is obvious that more overlap across different cognitive load levels exists for the glottal flow features.

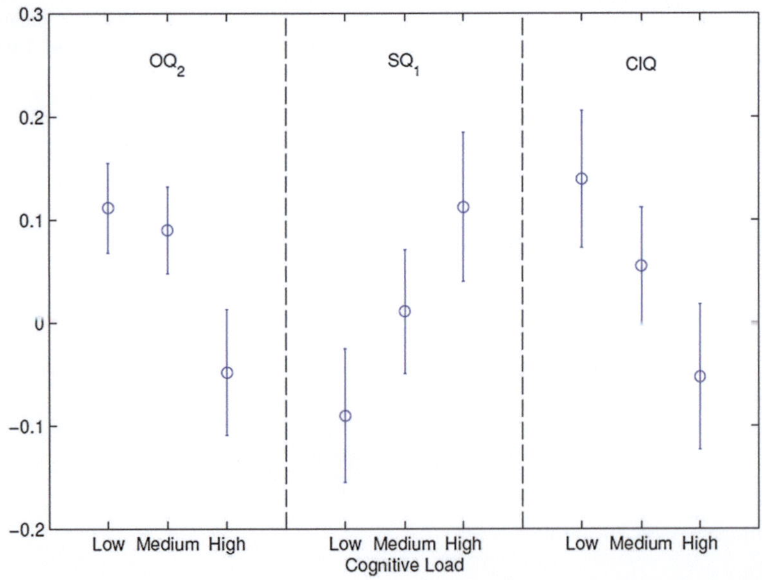

Fig. 7.9 Means and 95 % confidence intervals of three glottal flow features across different cognitive load levels

Table 7.5 3-class classification performance of glottal flow features

Features	Average accuracy (%)
OQ_1	39.8
OQ_2	46.8
QOQ	28.7
OQ_a	37.0
SQ_1	51.4
SQ_2	42.1
NAQ	44.9
AQ	39.8
ClQ	49.5

As a consequence, it is not surprising that the classification results of three cognitive load levels using the glottal flow features underperform the EGG features, as tabulated in Table 7.5.

7.5 Cognitive Load Classification System via Speech

High levels of consistency in training and testing speech data is necessary for statistical modelling and realistic cognitive load classification systems, however two major problems arise from the inconsistency of speech between speakers. The first problem is speaker variation, which implies that speakers may change their style of speaking due to different speaking content, time of speaking, mental status etc. This effect can be dealt with via feature warping [20]. The other problem relates to channel mismatch, which is normally caused by the short-term distortions, linear channel effects and other interferences, and can be reduced using Cepstral Mean Subtraction technique [21]. As suggested by Tet [14], a typical speech-based cognitive load classification system is illustrated in Fig. 7.10.

7.6 Summary

This chapter reviewed the cognitive load examination methods via speech. The typical experiments to induce different cognitive load levels were introduced, and features based on the source of speech and the filter, ie the modal of the vocal tract, are found to be applicable for cognitive load examination. The comparison between the features extracted from the source and the filter does not reveal a single obviously superlative feature, however the key insight here is that both sets of features are capable of achieving an overall cognitive load classification rate over 50 % for three levels of cognitive load. These studies not only provide direct insight

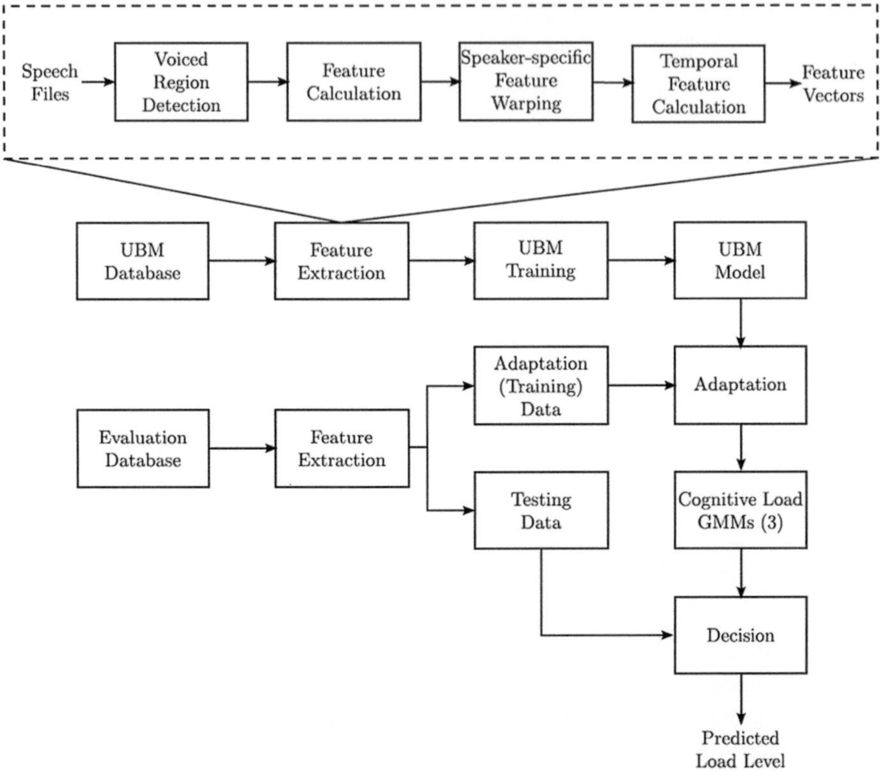

Fig. 7.10 A schematic representation of a speech-based cognitive load classification system [14]

into how cognitive load affects the glottal source and vocal tract, they can also be used to as guidelines for the design of new speech-based cognitive classification systems.

References

1. A. Berthold, A. Jameson, Interpreting symptoms of cognitive load in speech input., in *Seventh International Conference on User Modeling (UM99)*, 1999
2. C. Mueller, B. Grossmann-hutter, A. Jameson, R. Rummer, F. Wittig, Recognising time pressure and cognitive load on the basis of speech: An experimental study, in *Proceedings of the Eighth International Conference on User Modeling (UM 2001)*, Berlin, 2001
3. B. Yin, F. Chen, N. Ruiz, E. Ambikairajah, Speech-based cognitive load monitoring system, in *IEEE International Conference on Acoustics, Speech and Signal Processing, 2008. ICASSP 2008*, 2008, pp. 2041–2044
4. J.R. Deller, J.H.L. Hansen, J.G. Proakis, *Discrete-Time Processing of Speech Signals* (Institute of Electrical and Electronics Engineers, New York, 2000)

5. The Lexile Framework for Reading, *Matching Readers with Texts*. [Online]. Available: https://www.lexile.com/. Accessed 2 Feb 2015
6. T.F. Yap, J. Epps, E. Ambikairajah, E. Choi, An investigation of formant frequencies for cognitive load classification, in *Proceedings of Annual Conference of the International Speech Communication Association (InterSpeech'10)*, Makuhari, Japan, 2010
7. B. Yin, N. Ruiz, F. Chen, M.A. Khawaja, Automatic cognitive load detection from speech features, in *Australasian Computer-Human Interaction Conference (OzCHI'07)*, Adelaide, Australia, 2007, pp. 249–255
8. M. Daneman, P.A. Carpenter, Individual differences in working memory and reading. J. Verbal Learn. Verbal Behav. **19**(4), 450–466 (1980)
9. M.J. Kane, D.Z. Hambrick, S.W. Tuholski, O. Wilhelm, T.W. Payne, R.W. Engle, The generality of working memory capacity: A latent-variable approach to verbal and visuospatial memory span and reasoning. J. Exp. Psychol. Gen. **133**(2), 189–217 (2004)
10. A.R.A. Conway, M.J. Kane, M.F. Bunting, D.Z. Hambrick, O. Wilhelm, R.W. Engle, Working memory span tasks: A methodological review and user's guide. Psychon. Bull. Rev. **12**(5), 769–786 (2005)
11. D. Conway, I. Dick, Z. Li, Y. Wang, F. Chen, The effect of stress on cognitive load measurement, in *Human-Computer Interaction – INTERACT 2013*, ed. by P. Kotzé, G. Marsden, G. Lindgaard, J. Wesson, M. Winckler (Springer, Berlin/Heidelberg, 2013), pp. 659–666
12. T. David, A robust algorithm for pitch tracking (RAPT), in *Speech Codning and Synthesis* (Elsevier Science, Amsterdam, 1995), pp. 495–518
13. N.L. Phu, The use of spectral information in the development of novel techiques for speech-based cognitive load classification, University of New South Wales, 2011
14. Tet Fei Yap, Speech production under cognitive load: effects and classification, University of New South Wales, 2011
15. K.R. Scherer, D. Grandjean, T. Johnstone, G. Klasmeyer, T. Bänziger, J.H.L. Hansen, B. Pellom, Acoustic correlates of task load and stress|Affective Sciences
16. E. Mendoza, G. Carballo, Acoustic analysis of induced vocal stress by means of cognitive workload tasks. J. Voice **12**(3), 263–273 (1998)
17. G.R. Griffin, C.E. Williams, The effects of different levels of task complexity on three vocal measures. Aviat. Space Environ. Med. **58**(12), 1165–1170 (1987)
18. H. Boril, S. Omid Sadjadi, T. Kleinschmidt, J.H.L. Hansen, Analysis and detection of cognitive load and frustration in drivers' speech, in *Proceedings of INTERSPEECH 2010*, Makuhari Messe International Convention Complex, Chiba, Makuhari, Japan, 2010, pp. 502–505
19. M. Airas, TKK Aparat: An environment for voice inverse filtering and parameterization. Logoped. Phoniatr. Vocol. **33**(1), 49–64 (2008)
20. J. Pelecanos, S. Sridharan, Feature warping for robust speaker verification, in *Faculty of Built Environment and Engineering*, Crete, Greece, 2001, pp. 213–218
21. B.S. Atal, Effectiveness of linear prediction characteristics of the speech wave for automatic speaker identification and verification. J. Acoust. Soc. Am. **55**(6), 1304–1322 (1974)

Chapter 8
Pen Input Based Measures

Writing, as a complex form of interaction, often entails intensive user attention and cognitive load. This chapter discusses cognitive load examinations involving handwriting, with a focus on the behavioral features which involve writing velocity, pressure, pen gesture and some features derived from them. Based on the cognitive load examination of different scripts including text, digits and sketches, it is suggested that the behavioral features for cognitive load examination is content dependent and sometimes writing direction dependent. Specifically, this chapter discusses the following issues:

- Handwriting datasets for cognitive load examination, each focused on text writing, digit writing and sketching respectively;
- Writing features to examine cognitive load at stroke, sub-stroke or point levels;
- Implications of cognitive load on writing shape.

The discussed investigation methods to classify cognitive load levels are a first step towards a practical cognitive load aware writing system, and the preliminary results are likely to be further improved with more complex and refined methods.

8.1 Writing Based Measures

Writing behaviors involve the movement of the fingers and arm, the way strokes are generated and the pressure exerted on the pen. The first question posed is what the roles of the fingers and wrist play during writing. The biomechanical explanation for finger and arm movements is that each joint functions like a hinge or a universal joint. The hinge allows for movement control in one single direction, while the universal joint supports the movements of multiple directions. The wrist is a typical hinge and can only rotate a limited angle during writing. The fingers, however, are much more versatile as a combination of hinges and universal joints, with two or more degrees of freedom [1]. The difference in biomechanical properties between

F. Chen et al., *Robust Multimodal Cognitive Load Measurement*, Human–Computer Interaction Series, DOI 10.1007/978-3-319-31700-7_8

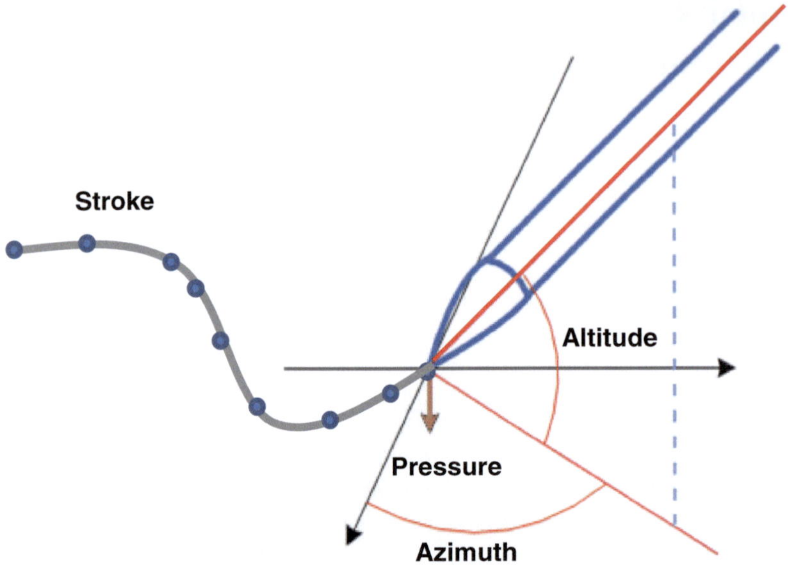

Fig. 8.1 Features available from digital pen writing

fingers and wrist determines their different functions. The fingers are responsible for the movement direction control of the pen and fine tuning of the written strokes, while the wrist mainly takes care of the writing behaviors perpendicular to the direction of forearm (Fig. 8.1).

Amongst the many behavioral features, writing velocity has been comprehensively investigated as this information is easily accessible from many writing tablets. Usually writing velocity is defined as the displacement of the pen tip within a time unit, for which only the period when the pen is writing is considered [2]. Previous studies have identified that writing velocity is affected by mood, such as sadness [3], depression [4] or other feelings of the writer [5]. Another factor that affects the writing velocity is the time limit of the writing task, and usually the writing velocity and the allowed writing time are inversely related.

Writing pressure indicates the force exerted on the pen tip against the paper when writing. It has been suggested that the writing pressure reflects personality perspectives including expressiveness, energy, determination, dominance and attitudes [6, 7]. In consequence, pressure-sensitive writing devices can be used for writer identification [8] or signature verification [9]. Also, writing pressure is affected by age, and studies have identified that children exerted less pressure in writing [10].

Newer writing devices, such as WACOM tablets, have made it possible to detect the orientation of the pen, and the pen orientation data reflects the way the pen is grasped, which forms another category of behavioral feature. The pen orientation feature is usually decided by two angular components, which describe the spatial tilt direction and amplitude of the pen. Due to its recent and limited availability on writing tablets, the orientation feature has not been intensively investigated for cognitive load research. However many applications have been proposed utilizing the pen orientation information as input commands, such as the Tilt menu [11] or other forms of tilt control [12].

A recent study by Gil Luria on digit writing tasks showed that writing behaviors are capable of reflecting cognitive load variations [13]. He examined the mean and standard deviations of writing duration, angular velocity of writing and pressure. The results, based on numerical writing with different gaps, showed that angular writing velocity is decreased, while writing duration is increased under high cognitive load configurations. However, in his study only digits were involved, and there was no quantity control on the digits that were written in the writing sample, so there is no further conclusion on what type of strokes or behaviors is more sensitive to cognitive load variations.

Another study conducted on CLTex [14] has revealed that writing pressure and velocity are good indicators of cognitive load variations. Specifically, a proportion of the writing points, delimited by the relative altitude span, are more sensitive to cognitive load variations compared with the rest of the writing points. The altitude span is based on the statistics of the points on a relative scale for individuals. So, when mapping the altitude span back to physical angle, it is also dependent on the usual writing altitude of the subject. As a consequence, for a specific subject, the span can be identified from the altitude distribution of his/her collection of writing points.

On the other hand, using the curvature to select writing data, together with writing velocity features, is able to discriminate cognitive load levels. According to the span-based investigations, all the spans perform similarly and no individual curvature span is especially sensitive to cognitive load variations. Although the result was not compared with Luria's study which suggests that the angular velocity changes with cognitive load levels [13], one finding that both studies can confirm is that writing curvature, together with writing velocity, is a cognitive load-sensitive factor to consider in a writing system.

8.2 Datasets for Writing-Based Cognitive Load Examination

Three datasets were specially collected, aiming to examine the relationship between cognitive load and different contents of writing. The later discussion on features and methods for cognitive load classification are largely based on the examination of the three datasets. The same device has been used to collect the three datasets, ie the WACOM DTZ-1200 W tablet. It samples at 142 Hz, and is capable of detecting the orientation of the pen within an azimuth range of 360° and an altitude range of 90°. The pen-tip can sense 1024 pressure levels, of which approximately 700 levels were typically recorded during the tasks.

8.2.1 CLTex Dataset

CLTex (Cognitive Load via Text) is a dataset composed of the writing samples of twenty subjects under three cognitive load levels. Several rules have been applied in the writing data collection. Firstly, the writing tasks should be able to induce several different levels of cognitive load effectively, which forms the primary goal of the data collection. Secondly, the tasks should be easily performed without learning required, which will avoid the learning effect and result in different cognitive load levels for the same type of task. Thirdly, the cognitive load effect should be induced during the writing process, or be experienced by the subjects exactly when they are writing, in order that the cognitive load changes are reflected in writing behaviors. Finally, the subjects should focus on the writing process, which means that the writing process should be as natural as possible.

The writing task for the CLTex dataset involves composing a sentence around specified words. The sentence composition task has low requirements on language and writing skills, which can be easily grasped by most educated subjects. It is also capable of inducing different levels of processing demand in writing via varying the number of given words. As a natural task, sentence composition can be conducted by subjects with a high level of engagement.

Specifically, in each writing task, a set of specified words were shown to the subjects, who were required to compose a sentence containing the given words, and write down the composed sentence. The words were selected based on their frequency of usage and considered to be easily understood by subjects, and only nouns, verbs, adjectives and adverbs were selected. The number of given words corresponds to the cognitive load levels: writing a sentence with one word was considered to induce low cognitive load, using two words to compose a sentence results in medium cognitive load, and a high cognitive load is expected to be experienced by the subject when a sentence is composed including three given words. Typical examples of the sentence composing tasks are shown in Fig. 8.2.

Eighteen college students aged from 21 to 32, including six females, participated in the experiment. All the subjects were frequent computer users (usage over 30 h per week). One left-handed subject was included in the research. Three of the subjects had used computer-based writing systems including mobile phones supporting stylus writing, while most subjects had experience in using touch

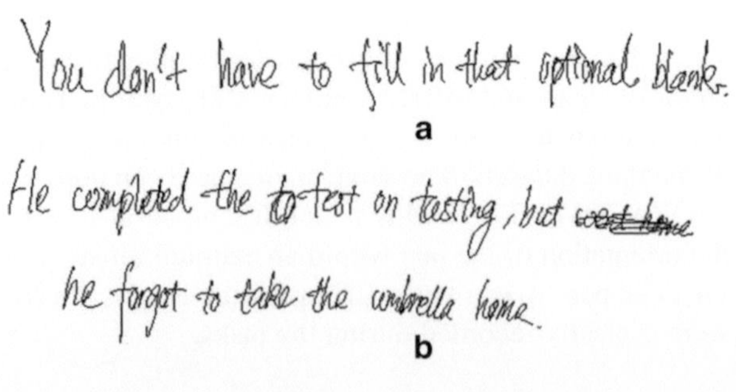

Fig. 8.2 Examples of writing with low cognitive load: (**a**) (given word: optional) and high cognitive load, (**b**) (given words: complete, forget, umbrella) during the CLTex task from the same subject

interfaces. Ethics approval was granted for this study, and each subject signed their consent before the test. The writing tablet was positioned parallel to the edge of the writing table by default, while two subjects requested to adjust the position of the writing tablet anti-clockwise before the experiment. After five training trials, each subject completed three blocks of fifteen trials each. Three other blocks of trials were also conducted to examine the effect of the digit sequence on the cognitive load. The blocks were randomized differently for different subjects to counteract the learning effect. All subjects finished the tasks within 45 min. For each block of tasks, we collected the subjective ratings on a 9-point scale (1: lowest difficulty, 9: highest difficulty), on which we believe the perceived difficulty will reflect the cognitive load experienced. During the experiment, a coordinator helped to initialize the testing environment, but did not assist the subject in completing the tasks. Afterwards, task feedback from the subjects was collected and recorded by the experiment coordinator on the task sheet.

8.2.2 CLSkt Dataset

Sketching constitutes an important perspective of pen interaction on writing interfaces, as any pen gestures that are used to operate a pen interface, for example, a tick to activating a command or a set of straight lines to unlock a screen, can be taken as typical forms of sketching. Furthermore, painting with digital pen, generally speaking, also constitutes sketching, and very small sketches can be considered as representative of prototypical characters. Understanding the cognitive load of sketching is an important step to improve the interface design and subjective experience. Hence, the CL Sketching (CLSkt) as the first sketching dataset for cognitive load examinations has been collected.

The general requirements of the CLSkt dataset are similar to those of the CLTex dataset: the subjects were intended to experience scalable cognitive load with full engagement during the writing process, and the tasks should be easy to comprehend to avoid learning effects or inducing extra cognitive load that is an artifact of a specific task design. Due to the nature of sketching tasks and limited prior knowledge on the performance evaluation methods for sketching, the construction of strokes and sub-strokes in the experiment design has been emphasized.

Specifically, as shown in Fig. 8.3a, digits were shown adjacent to the external edge of the dots for a few seconds (the time depending on the number of digits). The subject needed to memorize the location of the digits during this period. Then the digit indices disappeared, and the subject was requested to link the dots according to the ascending order of the indices. It should be noted that before the indices disappeared, the pen interface was not activated for sketching. There was no constraint on the sketching time, although the subjects were instructed to finish the task as soon as possible and a response time was recorded. Revising the sketches was allowed, by rewriting directly on the interface. However, pen traces other than

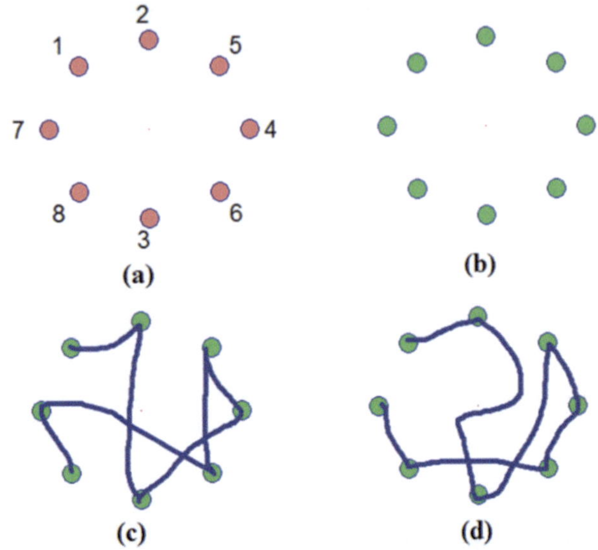

Fig. 8.3 Procedure of writing data collection in ClSkt: (**a**) a digit sequence is shown, (**b**) sketching interface after the digits disappear: each writer was requested to link the dots according to the ascending order of the digits, (**c**) correct sketch consistent with the ascending digit sequence, (**d**) incorrect sketch that violates the digit sequence

the linking actions were not allowed on the interface. The number of digits was utilized as the variable for manipulating the induced cognitive load. It is known that the length of the digit sequence could be a stimulus to vary cognitive load in a standard digit span test [15, 16], as longer digit sequences place a greater capacity requirement on the working memory, and thus increase the cognitive load. In consequence four digits (1–4) were adopted for the low cognitive load level, six digits (1–6) for medium levels and eight (1–8) for high levels. The digit sequences were randomized and manually examined to avoid extremely easy cases such as 1-2-3-4-5-6. The example in Fig. 8.3 is a typical task for writing under high level of cognitive load.

8.2.3 CLDgt Dataset

Cognitive Load via Digits (CLDgt) is a dataset constructed to investigate the impacts of cognitive load on handwriting scripts on the recognition performance. To construct CLDgt, the aim of the task design was to mentally load subjects' working memories when they execute writing tasks. Handwritten digits were preferred for four reasons. Firstly, digits share simple shapes and writing them is well practiced by most people, which reduces the factor of writing experience as a possible source of unwanted variability. Secondly, individual unconnected digits can be meaningfully elicited, so that per-character analysis on a large scale is possible. Thirdly, digits resemble some Latin characters in terms of stroke shape and sequence, which makes it possible to generalize our findings somewhat to text writing. Finally, digit tasks, such as N-back test and arithmetic calculation, are well-understood methods to

Table 8.1 Calculation tasks to induce three cognitive load levels

Cognitive load level	Digit	Carry	Example
Low	Single	<=1	3+4−2
Medium	Double	2	41−29+38
High	Three	3	365−277+409

conveniently induce many cognitive load levels. Specifically, N-back span tasks and calculation tasks were adopted as shown in Table 8.1.

The N-back tasks were conducted over 2 days, and three sessions were completed each day by each subject. 200 N-back tasks were involved in each session, where N was set to 0, 1 and 3 respectively, so in total 1200 N-back tasks were completed by each writer. Similarly, the calculation tasks were conducted over 2 days, comprising three sessions each. Each session involved 60 calculation tasks, and a total of 360 calculation tasks were conducted by each subject. The N-back and calculation sessions were randomized on each day, and each subject spent in total approximately 3 h on the tasks excluding between-session rests. The experiment time was comparatively long, as many repetitions of a digit were required for the digit-dependent analysis. As a result, not many subjects were involved in this experiment.

For both the N-back and the calculation tests, the occurrences of digits were balanced, meaning that if the subjects answered all the answers correctly, there would be an equal number of samples of each digit from 0 to 9, ie 120 samples for each digit from 0 to 9 for the N-back tasks, and 72 samples for each digit from 0 to 9 for the calculation tasks.

8.3 Stroke-, Substroke- and Point-Level Features

According to the different structural properties, eg writing direction, curvature, position etc., one stroke can be segmented into different parts, for example, straight line, curve, corner [17], each corresponding to one or more categories of hand behavior. The fine segmentation of written strokes makes it possible to investigate writing behaviors at different parts of a single stroke. Most methods discussed in this book employ spatial and temporal related information to segment strokes for writing feature investigation, however as stroke segmentation is not the focus of this book, the corresponding details will not be discussed here.

Decomposing strokes into structural components helps to investigate the strokes at microscopic levels, while the module-based methods aim to understand handwriting from a macroscopic perspective. Different modules are considered responsible for handwriting production [18], and each module refers to the sub-movement controller which is well trained during the learning of writing. For example, to write a pen trace, three modules may be involved, including timing accuracy, timing speed and force-amplitude accuracy [19]. An earlier study by Bernstein supports the module-based writing explanation, with the observation that writing at different sizes, or changing the writing limbs do not change the spatial or temporal accuracy of the written text [20].

Behavioral features can be extracted from the characters, strokes, or even every single sample point within one stroke. Behavioral features are classified into two categories, ie spatial features and temporal features. Spatial features describe the physical location of the pen strokes, and can be partially retrieved from the written strokes. Typical spatial features include local physical maximum and minimum of the strokes, stroke length, stroke height and width. Spatial features are promising in deciding the structure of the written scripts, but carry limited information on the dynamics of the writing hand. In contrast, temporal features indicate the physical movements of the writing hand, which include the writing velocity, pressure, and the way the pen is gripped. Practically, temporal features carry more information than spatial features, as the spatial information can be retrieved from a comprehensive collection of temporal writing information (Fig. 8.4).

According to a recent investigation on the CLSkt dataset [14], stroke-level classification rate was better than that of the sub-stroke level, however new insights should be noted regarding the sub-stroke level cognitive load classification: a further investigation of the sub-strokes as a function of different writing directions suggests that horizontal sub-strokes are more reflective of cognitive load variations than the strokes writing in other directions, and the best classification rate for three cognitive load levels was achieved at 62.6 %, and two levels at 79 %. It has been identified as a general principle that hand movements during writing can be resolved into two perpendicular directions [14], and generally the finger movements dominate the writing of horizontal sub-strokes, while the vertical sub-strokes are closely related to the wrist movements. Sub-strokes written in the other directions comprise combined movements of both the fingers and the wrist. It was proposed that cognitive load variations are well reflected in finger movements,

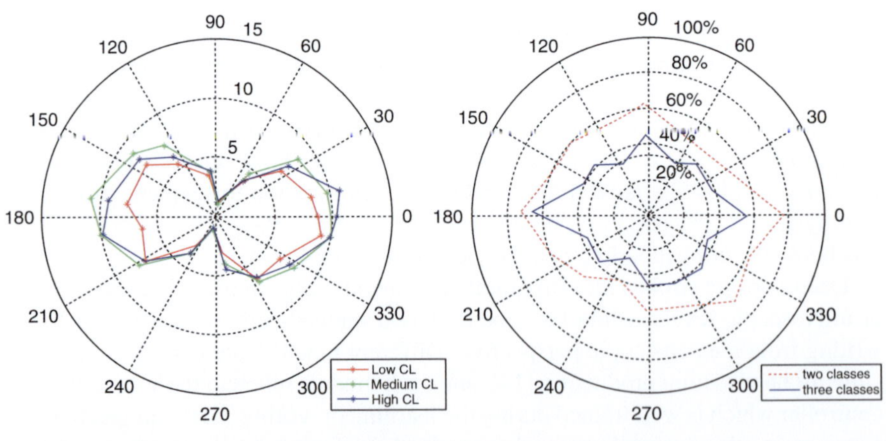

Fig. 8.4 Direction-based cognitive load analysis: (**a**) Velocity distributions in 16 directions of 3 cognitive load levels, (**b**) classification accuracy for 2 and 3 cognitive load levels in 16 directions using mean velocity as feature

and a possible cause is that compared with the wrist movements, the finger movements are a more complex biomechanical action and more capable of adaptation to the changing task requirement. Evidence have been found that the writing velocity is higher horizontally than vertically, which implies a higher range of velocity changes from the finger than from the wrist. In consequence, with increased cognitive load, increasing the movement of the fingers is a natural response to produce a higher writing velocity. It has not been not found that the same trend exists for wrist movements in the experiments, however it is still possible that the movement patterns of the wrist may be altered when extremely high cognitive load is experienced by the subject, and increasing the finger movement alone is insufficient to release the selective attention for conducting the writing task.

8.4 Cognitive Load Implications on Writing Shapes

As we have shown that a writer's cognitive load affects the writing shape, and thus has implications for the recognition performance of digits. This finding has important implications for pen system design, since a critical criterion for a user-friendly pen system is its capability to interpret the writer's intention correctly, and provide a precise response to the writer's pen input. It is postulated that a cognitive load sensitive writing system will be able to achieve higher recognition accuracy. Practically, the knowledge of the writer's behaviors under different cognitive load conditions will help to customize the handwriting input systems and improve their recognition performance.

In most situations, existing writing interfaces neither detect nor report the cognitive load of the writer, nor utilize the cognitive load information, which could be deduced from the context, to assist the recognition of written characters. Therefore neglecting the cognitive load conditions of the writers is likely to result in a poorer user experience. For example, when a writer is experiencing high cognitive load and focusing most mental resources on the content of writing, frequent occurrences of handwriting recognition errors will force the writer to split attention to improve the writing shape, which inevitably competes for mental resources with the primary task of content composition, and the frequent switches of attention between writing content and writing behavior is likely to result in a vicious circle that will ultimately increase the cognitive load and degrade the user experience. However these consequences might be avoided if the recognition system is capable of processing the deformed characters with improved accuracy based on the perceived cognitive load of the writer.

Here we demonstrate that cognitive load affects the writing behaviors by degrading the written shape during N-back and calculation tasks. This recognition-based study confirmed the hypothesized relationship between cognitive load and recognition performance: as cognitive load increases, the recognition accuracy decreases for most digits and thus degrades the overall recognition

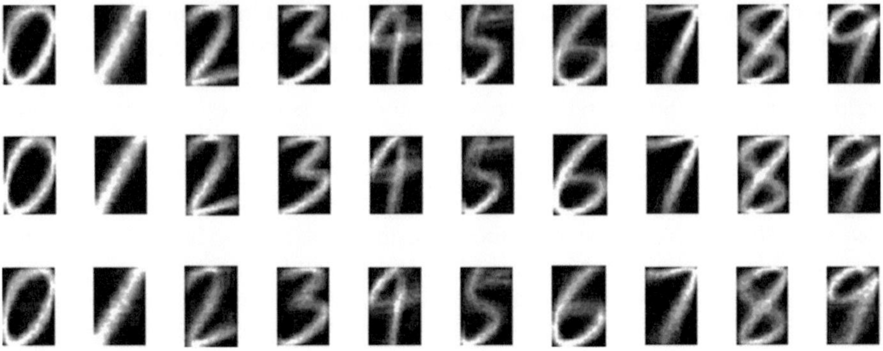

Fig. 8.5 Pixel probability distribution reflecting the stroke shape variations of written digits from one writer for three recall tasks. The first row corresponds to low cognitive load (0-back recall task), the second row to moderate cognitive load (1-back recall task), and the last row to high cognitive load (3-back recall task) [14]

performance, and a decrease of recognition accuracy of over 10 % was observed between low and high cognitive load conditions. As visualized in Fig. 8.5, this is due to the fact that writers have less control of the pen as they are writing. The examination with entropy measurements also demonstrated the increased distortions existent in the pen traces as higher cognitive load was experienced by writers. This artifact might be caused by the fact that under high cognitive load, the available mental resources allocated to manage the writing behaviors decrease. Although comparable results were not found from handwriting research due to the lack of similar studies, the impact of cognitive load on gait [21] was consistent with our finding, suggesting that the variations in stride length increase under high cognitive load. However, investigation of cognitive load effects on body posture control [22] has shown that under cognitive load conditions, decreased variance in body sway was observed compared with normal conditions, which was in conflict with our finding, possibly due to the dual-task protocol and involvement of external stimulation in their experiment.

Addressing the question of whether the writing of every digit is uniformly affected by the cognitive load, according to the investigation of individual digits, the answer is no. For digits composed largely of curvatures, including 2, 3, 6, 8, 9, their shapes are more sensitive to the different cognitive load conditions. In comparison, digits composed with straight lines, for example, 1, 4 and 7, seem to be more robust to the varied cognitive load. A possible explanation is that when writing a single digit, fewer changes in writing directions means that there is less potential for impact on writing by the cognitive load. This could also be interpreted as frequent changes of writing directions involve more fine adjustments of motor control, which competes for mental resources with any other parallel cognitive tasks, and this competition will result in unstable mental resource requirement and thus affect writing shape.

Interestingly, the matched training and testing set did not produce the best recognition rate. The copied digits from the benchmark writing set performed well in most cases, except when recognizing the digits written at the high cognitive load conditions. In contrast, the recognizer trained with moderate cognitive load performed better on the high cognitive load set, both for the calculation and for the N-back tasks. This finding implies that for the normal or close to normal writing when the writer does not experience high cognitive load, the writing samples collected under similar mental conditions are a sufficient training requirement and good recognition results are possible. However, for writing conditions under cognitive load, writing shape variations involved in the training set may resemble the deformed strokes in the recognition phase, and thus the recognition performance can be improved. The extent of distortion should not exceed the target cognitive load level, otherwise the deformed shapes will also bias the recognizer, which may be why training and testing with the matched data sets at the highest cognitive load condition did not get a better result.

8.5 Cognitive Load Classification System

Based on the above discussions, we are aware that cognitive load has implications on both the writing experience and the performance of writers, and thus we suggest that it is important to include cognitive load as a factor in the design of a writing system.

Since writing pressure, writing velocity and pen orientation features are capable of identifying the different cognitive load levels, it is desirable that the writing devices are able to capture these writing signals, which can be essential for writer cognitive load detection.

Based on the many cognitive load examination methods discussed above, different cognitive load aware writing systems can be constructed. One typical cognitive load classification system is proposed in Fig. 8.6, which takes advantage of the sensitivity of sub-stroke direction to cognitive load variations. Specifically, the features were only extracted from sub-strokes written in the horizontal direction, aiming to improve the classification accuracy and decreasing the computational cost.

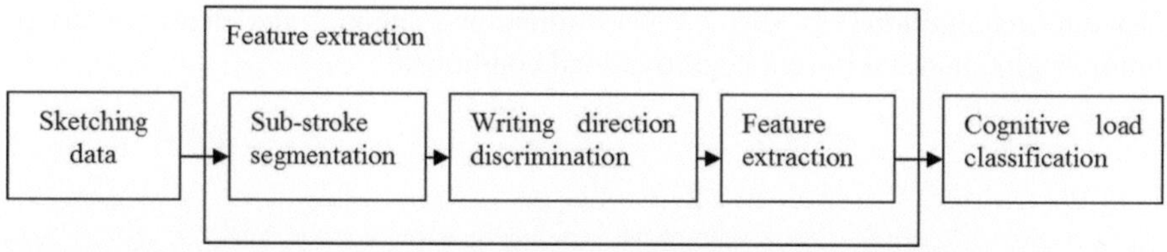

Fig. 8.6 Structure of a cognitive load classification system based on sketch data

This cognitive load classification system extends the methods of stroke-based cognitive load investigation to the sub-stroke level, by accounting for sub-stroke direction information. This constitutes two advantages compared with the stroke-based method: firstly, one stroke can be segmented into more sub-strokes, and hence more features can be extracted and more reliable system performance can be expected; secondly, the sub-strokes in different directions can be examined respectively, and only those sub-strokes written in the horizontal direction will be selected to extract features for cognitive load classification, because these are capable of achieving higher classification rate. Sub-strokes written in other directions can be excluded from further analysis and thus the total computational cost can be reduced.

From a pragmatic perspective, a cognitive load classification module can be easily embedded into the current pen systems and provide the writer's cognitive load information, which would be helpful in user experience enhancement and interface design. From the design perspective, the sketch-based cognitive load classification method can provide cues for interaction designers in interaction paradigm adaptation, sketching task selection and graphic interface decisions. The understanding of a writer's cognitive load can also benefit education, for example teaching painting – as the cognitive load experienced when young children learn to paint is difficult to quantify via performance (like the correct rate of calculation tasks).

8.6 Summary

This chapter introduced methods to measure cognitive load with writing and pen-based features. Essentially most writing behaviors, including writing velocity, pressure, and the gesture to grasp the pen can be reflective of cognitive load variations of the writer.

Decomposing the written shapes into different components, eg sub-strokes helps to improve the cognitive load classification performance, and sub-strokes written in different directions are sensitive to different extent when cognitive load is changed: specifically, the horizontal strokes are more suitable for cognitive load discrimination than the vertical strokes.

Cognitive load has implications on the creation of shapes via a pen, and a high cognitive load decreases the control of hand and thus increases the variation in the shape of written characters. This knowledge is important in improving current character recognition performance, via training and adapting the character recognition engine under different cognitive load conditions.

References

1. H.-L. Teulings, A.J.W.M. Thomassen, G.P. van Galen, Invariants in handwriting: The information contained in a motor program, in *Advances in Psychology, vol. 37* (Elsevier, London, 1986), pp. 305–315
2. X. Cao, S. Zhai, Modeling human performance of pen stroke gestures. Chi. Conf. **2**, 1495 (2007)
3. E.T. Higgins, Self-discrepancy: A theory relating self and affect. Psychol. Rev. **94**(3), 319–340 (1987)
4. W.D. Hale, B.R. Strickland, Induction of mood states and their effect on cognitive and social behaviors. J. Consult. Clin. Psychol. **44**(1), 155–155 (1976)
5. S.D. Mayes, S.L. Calhoun, Learning, attention, writing, and processing speed in typical children and children with ADHD, autism, anxiety, depression, and oppositional-defiant disorder. Child Neuropsychol. **13**(6), 469–493 (2007)
6. R.J. Klimoski, A. Rafaeli, Inferring personal qualities through handwriting analysis. J. Occup. Psychol. **56**(3), 191–202 (1983)
7. G.R. Pascal, Handwriting pressure: Its measurement and significance*. J. Pers. **11**(3), 235–254 (1943)
8. K. Yu, Y. Wang, T. Tan, Writer identification using dynamic features, in *Biometric Authentication, vol. 3072*, ed. by D. Zhang, A.K. Jain (Springer, Berlin/Heidelberg, 2004), pp. 512–518
9. R. Plamondon, G. Lorette, Automatic signature verification and writer identification – the state of the art. Pattern Recogn. **22**(2), 107–131 (1989)
10. T.L. Harris, G.L. Rarick, The relationship between handwriting pressure and legibility of handwriting in children and adolescents. J. Exp. Educ. **28**(1), 65–84 (1959)
11. F. Tian, L. Xu, H. Wang, X. Zhang, Y. Liu, V. Setlur, G. Dai, Tilt menu: Using the 3D orientation information of pen devices to extend the selection capability of pen-based user interfaces. Chi. Conf. **2**, 1371 (2008)
12. M. Rahman, S. Gustafson, P. Irani, S. Subramanian, Tilt techniques: Investigating the dexterity of wrist-based input. Chi. Conf. **2**, 1943 (2009)
13. G. Luria, S. Rosenblum, A computerized multidimensional measurement of mental workload via handwriting analysis. Behav. Res. Methods **44**(2), 575–586 (2012)
14. Kun Yu, Cognitive load examination via pen interactions, University of New South Wales, 2015
15. P. Barrouillet, S. Bernardin, V. Camos, Time constraints and resource sharing in adults' working memory spans. J. Exp. Psychol. Gen. **133**(1), 83–100 (2004)
16. P. Barrouillet, S. Bernardin, S. Portrat, E. Vergauwe, V. Camos, Time and cognitive load in working memory. J. Exp. Psychol. Learn. Mem. Cogn. **33**(3), 570–585 (2007)
17. C.M. Privitera, R. Plamondon, A system for scanning and segmenting cursively handwritten words into basic strokes, 1995, vol. 2, pp. 1047–1050
18. J.J. Schillings, R.G.J. Meulenbroek, A.J.W.M. Thomassen, Decomposing trajectory modifications: Pen-tip versus joint kinematics, *Handwriting and Drawing Research: Basic and Applied Issues*, pp. 71–85, 1996
19. S.W. Keele, H.L. Hawkins, Explorations of individual differences relevant to high level skill. J. Mot. Behav. **14**(1), 3–23 (1982)
20. D. Ilngle, The co-ordination and regulation of movements. Papers translated from Russian and German. N. Bernstein. Pergamon, New York, 1967. xii + 196 pp., illus. $8, Science, vol. 159, no. 3813, pp. 415–416, 1968
21. R.B. Eladio Martin, Analysis of the effect of cognitive load on gait with off-the-shelf accelerometers, 2011
22. G. Andersson, J. Hagman, R. Talianzadeh, A. Svedberg, H.C. Larsen, Effect of cognitive load on postural control. Brain Res. Bull. **58**(1), 135–139 (2002)

Chapter 9
Mouse Based Measures

Multimodal behavior and interaction have been shown as viable indicators of cognitive load, with strong evidence presented for the effectiveness of speech, interactive gesture and digital pen signals [1]. One key component of this multimodal behavioral model is user/mouse interaction. This chapter demonstrates the relevance of mouse interactivity based features as an indicator of a user's cognitive load. This chapter focuses on the following aspects:

- Demonstrating basics of user mouse interaction;
- Presenting the temporal and spatial mouse features that are found to be viable indicators of cognitive load;
- Assessing the possibility of incorporating the mouse interactivity features in multimodal cognitive load measures.

9.1 User Mouse Activity

Mouse activity is an important part in human-computer interaction. Though we acknowledge the existence of non-mouse based scenarios, yet most applications require some constant navigational component. This component can be used for both within task activity and also inter application switching. Using a handy device as mouse pointer is convenient. An alternate would be to manage application control via keyboard but few can manage this efficiently. Also the mouse pointing device is conveniently non-intrusive and does not require any special data collecting equipment. A regular stream of data points is obtained whenever mouse is operated.

Though mouse dynamics is generally being explored as a biometric technology, this chapter shows how this approach can be adapted and enhanced to detect pattern changes in user behaviour as cognitive load is varied. Mouse dynamics has been used in behavioural biometrics for both static and continuous authentication [2]. The current research was inspired by techniques used for continuous

F. Chen et al., *Robust Multimodal Cognitive Load Measurement*, Human–Computer Interaction Series, DOI 10.1007/978-3-319-31700-7_9

authentication as the problem environment bears a strong resemblance to that of continuous cognitive load monitoring. In our experiments mouse activity is recorded via Java AWT mouse listener. Interrupts are generated whenever the mouse is operated and relevant mouse events (moved, clicked, dragged etc.) recorded along with time stamp, screen coordinates (X and Y) and other application widget information.

9.2 Mouse Features for Cognitive Load Change Detection

Both temporal and spatial features are discussed in this chapter. The temporal features refer to the time dimension of the mouse activity signals. Using the time stamps associated with signals, the time difference (or 'pause') between consecutive signals or features can be calculated. We also discuss spatial (or trajectory based) features used for real-time cognitive load detection scenarios (discussed in Chap. 15).

9.2.1 Temporal Features

In a multitier experiment designed to study effects of cognitive load on trust [3], this section measures various behavioural modalities including mouse interactivity at two different cognitive load levels (high and low). Cognitive load was varied using a standard dual-task design [4]. The low load scenario contained only a primary task, whereas the high load condition included a secondary task as well, that periodically popped-up into the user's view and sometimes required a classification action otherwise the pop-up had to be ignored. Mouse events were recorded from 88 subjects, each of which completed two different tasks (labelled T1 & T3) twice (under randomized order of high and low cognitive load levels). These events were recorded with a precision of milliseconds. The most frequent time intervals (between two consecutive mouse events) were zero ('operationally labelled' as anything less than 15 milliseconds), 15 milliseconds and 16 milliseconds. Other values ranged from multiples of this least count (in milliseconds) to tens of seconds. Only non-zero time intervals were considered for pause (temporal) analysis. It was decided to differentiate between time intervals greater than 1 s (now called Contemplation-style pause) and those less than 1 s (now called Hesitant-style pause). Any 'pause' greater than one-tenth of a second and less than 1 s apart was considered to be part of hesitant style contiguous activity. Pauses greater than 5 s were ignored due to low count and also as they are more difficult to interpret due to additional latent factors.

In general, a 'pause/break' in pointing device activity could be due to: (a) an external distraction that interrupted user from the current task at hand, (b) user switching focus to other input device eg keyboard, or (c) the user taking moment/s to think situation over (indicating high working memory usage). In this case, situation (a) was eliminated by close monitoring and controlled experimental conditions that avoided external distractions. Situation (b) was avoided by careful selection of tasks that relied solely on mouse input device. This allowed situation (c) to be the most probable case in the experiment.

Typical interactive computer sessions are comprised of varied duration sequences of either user intently observing/studying screen with no input activity (these may be termed as 'inactive' phase of user interaction) or user observing/monitoring screen accompanied with rapid spurts of input activity (this may be termed as 'active' phase of user interaction). Note that the terms 'active/inactive' are used here with respect to 'input' dimension of user interactivity.

Mouse events were monitored continuously throughout the experimental session including both active and inactive user interaction phases. The active phase was typically characterized by mouse events less than a second apart. Whereas the inactive phase had mouse events more than 1 s apart. The former was referred as *Hesitant-style pause* and the latter as *Contemplation-style pause*. Consecutive mouse events during active phase were closely spaced (typically less than 1 s) and the source of variation in this distribution can be associated with hesitant or cautious behavior of the user due to high load on working memory. While mouse events more than 1 s apart can be attributed to user pausing to contemplate the next move or recovering from lag due to task switching. Both behaviors are indicative of high load on working memory.

Paired-*t* test (with 95 % confidence interval) was performed for 88 subjects, separately, for Contemplation-style and Hesitation-style Pause frequency. Both Task 1 and Task 3 demonstrated significant p-values for Contemplation-style pause and Hesitant-style pause as well. Due to the longer duration of Task 1, overall pause frequency recorded (in both types) was higher with more pronounced differences. Task 3 was relatively a shorter task with low pause frequency. However, in each case, the pause frequency was significantly different between the low and high load scenarios. Detailed distribution (using binned bar chart) of each style pause is presented in Figs. 9.1 and 9.2 for trend comparison. Overall there appeared to be an increasing difference in pause frequency between the two load conditions with increasing pause event frequency.

Preliminary investigations of pause duration and frequency allowed us to differentiate between larger and relatively smaller duration pauses. Larger duration pauses were identified mostly by patches of inactivity and considered as contemplation-style pauses. The observed patterns of contemplation style pauses in mouse behavior resembled the previously observed patterns of speech pause features, where it was observed that people use more and longer pauses (both silent and filled) under high cognitive load conditions versus low load conditions [5]. The pattern of more frequent pauses for high cognitive load scenarios is demonstrated consistently and is evident from the binned contemplation pause averages in Fig. 9.1. The bins for 5 s and over were ignored due to low count and more complicated latent factors. The mean difference between contemplation type pause was about 29 more for high cognitive load condition. Since these pause

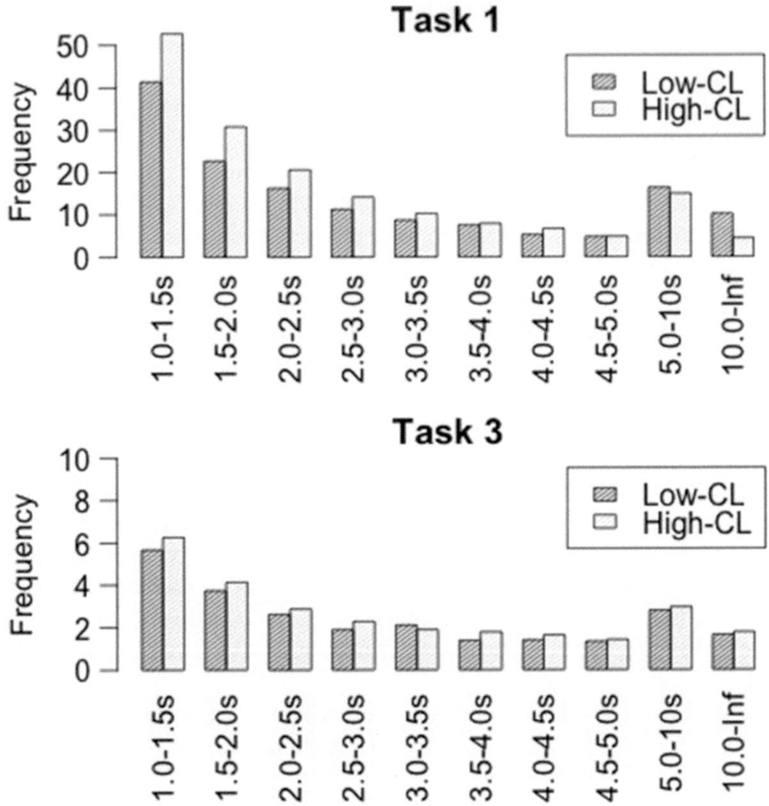

Fig. 9.1 Binned contemplation pause average

frequency difference values did not correlate to the number of times secondary task popped-up during high load condition, we attribute this difference to high load condition on user's working memory.

The relatively smaller duration pauses were mostly due to hesitant/thoughtful slow mouse movements by the user. These were labeled as hesitant-style pause. Hesitant-style pauses are more difficult to interpret directly and associate with high cognitive load. One simpler interpretation would be that increase in hesitant-style pauses is due to 'slow and cautious behavior' in performing primary task while anticipating the popping-up of secondary task. Binned hesitation-style pause averages (see Fig. 9.2) also show an increasing difference trend with increasing pause frequency.

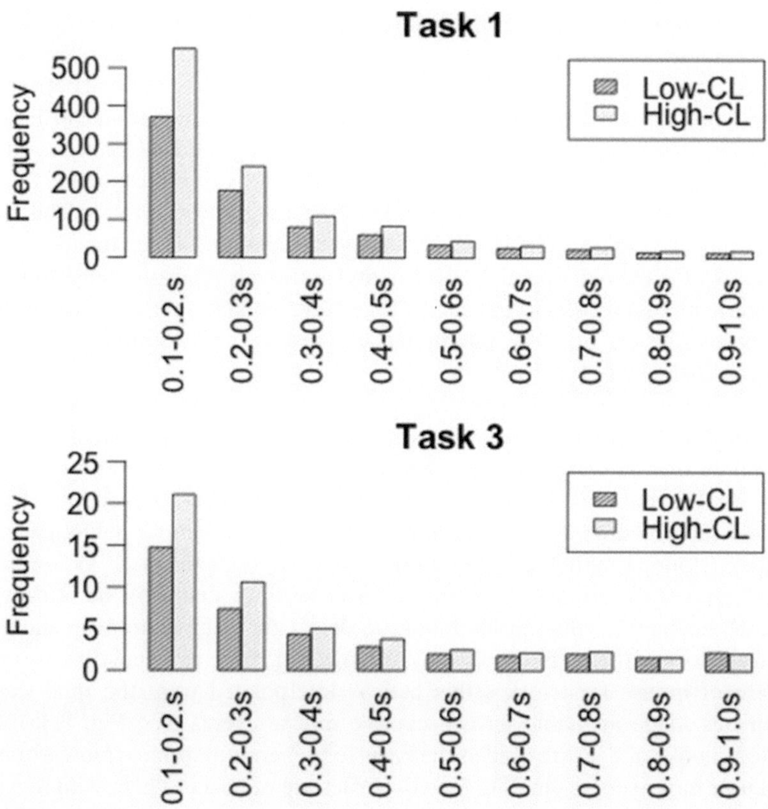

Fig. 9.2 Binned hesitation pause average

9.2.2 *Spatial Features*

With regards to spatial mouse features, much research has been carried out in the behavioural biometric research area [2]. Our spatial mouse feature research, in addition to our pause feature findings [3], was inspired by trajectory analysis techniques being investigated for user's continuous authentication [6]. Schulz [6] used mouse curve (ie mouse movements with little or no pause between them) features (eg length, curvature and inflection) to develop classification histograms that would be representative of an individual's typical mouse usage. This problem environment bears strong resemblance to that of continuous cognitive load monitoring scenario. However, the key problem, for biometric research, remains to minimize and handle intelligently the effects of within-user behaviour variability (for extended periods of mouse usage). Luckily, this presented an opportunity for research as it was already presumed that changes will be observed in user behaviour under varying cognitive load conditions – and behavioural biometric findings supported this case.

Raw mouse events (eg Move, Click, Drag etc.) are interrupt-driven and can be easily monitored. These events along with system time stamps and screen coordinates can be recorded. A 'pause' in mouse activity refers to the interval between two consecutive mouse events. It has been argued that both contemplation and hesitation style pause interval categories hold significant promise for detecting high cognitive load on working memory [3]. Contemplation-style intervals correspond to the more clearly observable break in user input activity that ranges from 1 to 5 s. This type of interval shows change patterns similar to those previously observed for speech pause features [7]. On the other hand, hesitation-style intervals typically range from more than one-tenth of a second to 1 s apart. These correspond to subtle variations in user input behaviour that may be interpreted as 'hesitant' or 'cautious' due to high cognitive load on working memory.

We extended this approach by extracting 'trajectories' associated with hesitation-style intervals, which moves from the temporal to the spatial dimension. User mouse activity was segmented into trajectory curves similar to those proposed by Schulz [6], but with more detailed features. The eight mouse curve features observed (and recorded) include length and number of sample points per curve; curvature (four components) and inflection (two components). These 'mouse curves' represent normalized chunks of user activity that may be streamed to real-time machine learning instantiations to detect changing patterns or anomalies.

A trajectory is the path followed by an object in space as a function of time. In the case of mouse trajectories, this path is interpolated from the time stamped coordinates made available by consecutive mouse events (see Fig. 9.3a). Time sampled chunks of data are used as the basis for generating splines (representing the actual user mouse curves in 2D). Several issues are of concern here with regards to 'mouse curves'; as to how closely the interpolated splines actually represent the user's behavior. Some of the extracted features (like actual location and velocity) are direct representation of user behavior, whereas others (like curvature and inflection points) are calculated approximations. From an operations point of view, the system continues to collect consecutive ('mouse-moved' type) raw data points (including screen X and Y coordinates) till a time-stamp difference of greater than 0.1 s and less than 1 s signals the completion of one mouse 'curve'. This bunch of raw observations (typically ranging from 10 to 200 points) is grouped to create one mouse 'curve'. From this 'curve', eight features (ie number of observation points, Euclidian length of curve, positive and negative curvatures of vertical and horizontal activity components, inflexion points for both vertical and horizontal movement) are extracted into a feature vector. Several contiguous feature vectors are passed on to decision engine as feature windows.

9.2.2.1 Spatial Features with Straight Lines

In an experiment study, the mouse data collected during the text-chat sessions (using instant messaging) between 20 participants were divided into 10 pairs. These data sets are the same as those used in [8], and were analysed and used to

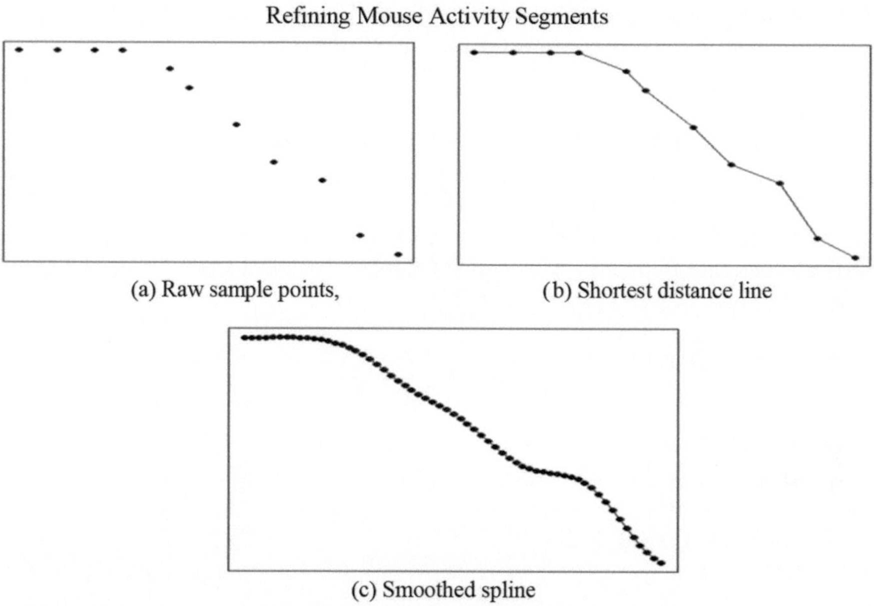

Fig. 9.3 Estimating user behavior from raw sample points

differentiate cognitive load level using mouse measurements. This section examines the straight lines in the mouse movements (vertical and horizontal lines).

The participants were asked to play a game known as DayTrader [9, 10] which involves two main tasks: investment and chatting. Each participant chatted with their partner for 30 min, divided into six sessions, so the length of each session was 5 min. This investment game was followed by social dilemma strategies where there was a high payoff for each amount of money invested. Each participant had to invest five times after each chat session and each participant could invest up to sixty dollar each time. Whenever a participant invested an amount, he/she then received a payoff (\$A), calculated as follows:

$$\$A = ((amount\ invested\ +\ partner\text{'}s\ amount\ invested)*3)/2$$

and the payoff for the money which was not invested (\$B) was calculated as follows:

$$\$B\ = (amount\ not\ invested\ *2)$$

This experiment included two conditions of cognitive load, categorized as low and high. When the participants were chatting, eight random numbers were shown on the computer screen which the participants were required to sum. For three chat sessions, the numbers were small; either one or two, this being the low cognitive load condition. For the other three chat sessions, the numbers were between 100 and

300, this being the high cognitive load condition. For each cognitive load condition, the participants were asked to rank the mental effort required to complete the task of summing the numbers, adapted from [11], to validate the cognitive load level induced.

The mouse coordinates (*X* and *Y*) were captured and used to observe mouse movement. The number of straight lines between each sequential pair of coordinates (*X* and *Y*) was calculated. Specifically, by using the coordinates, the total number of horizontal lines (the frequency of zero slopes) and the total number of vertical lines (the frequency of undefined slopes) was calculated. To calculate the slopes, the follows formula was used [12]:

$$Slope = \frac{Y2 - Y1}{X2 - X1}$$

where $X1$ and $Y1$ are the first pair of coordinates and $X2$ and $Y2$ are the second pair of coordinates. For the zero slope, both $Y1$ and $Y2$ have the same value $(0/(X2-X1) = 0)$ while for the undefined slope, both $X1$ and $X2$ have the same value $((Y2-Y1)/0 =$ undefined result$)$ [12].

9.2.2.2 Straight Line Results

A two-tailed *t*-test was used to compare the answers to the question about the level of cognitive load and the number of straight lines between participants over the different cognitive load conditions. There was a significant difference in the ranking of the participants in relation to the mental effort required in the different cognitive load conditions ($p < .000$) [8]. For the high cognitive load, the mean of the ranking of the mental effort was 7.25 while for the low cognitive load, the mean of the ranking of the mental effort was 2.5 [8] (see Fig. 9.4), thus validating the differential load induced by the task.

The number of straight lines between the *X* and *Y* coordinates on the graphical user interface were counted under the two cognitive load conditions. The findings show that the participants moved the mouse differently when under the different load conditions. Specifically, the findings show that the participants moved the mouse horizontally significantly more frequently ($p < .01$, Fig. 9.5a) when they were under a low cognitive load where the total number of the horizontal lines was 5347 while under the high cognitive load, the total number was 3227. Also, the finding for the vertical lines ($p = < .05$, Fig. 9.5b) were similar to the findings for the horizontal lines, with a total number of 5314 for the low cognitive load level and a total number of 3686 for the high cognitive load level.

Fig. 9.4 The average of ranking the mental effort level (using a Likert scale from 1 to 9 where 1 indicates a low effort and 9 a high effort)

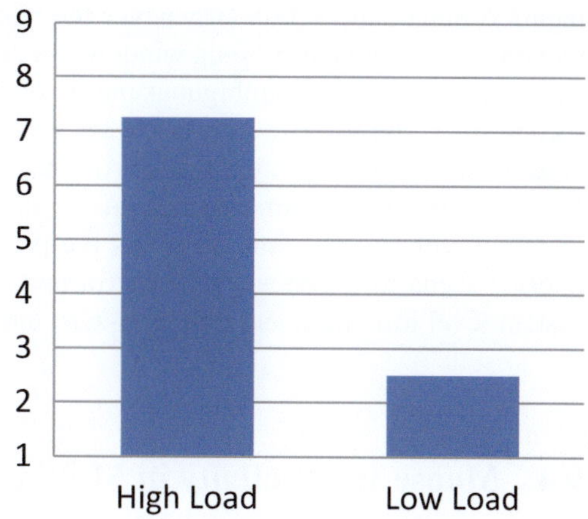

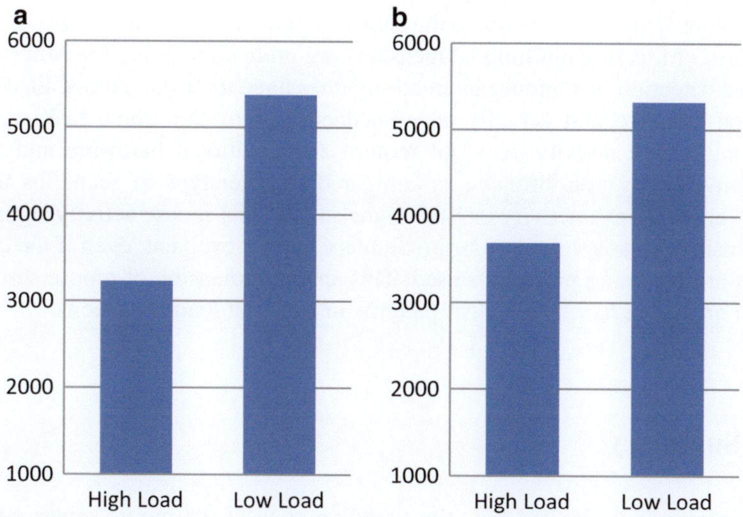

Fig. 9.5 Mouse movement features: (a) The total number of horizontal lines, (b) The total number of vertical lines

9.3 Limitations of Mouse Feature Measurements

The normal course of user action would be to perform primary task related activity at a reasonable pace resulting in relatively uniform mouse movements (typically events less than a second apart) punctuated with reasonable inactive patches that may be interpreted as contemplation style pauses. However, with introduction of secondary task, the user is required to divert attention towards the fixed location of a screen pop-up and then, based on pop-up message category, decide whether to respond (by clicking) or simply ignore. Since the pop-up was time sensitive and required immediate action (whenever the relevant category shows up) the mouse movements towards the pop-up window were typically a jerk (less than one-tenth of a second apart). Due to ambiguous nature of mouse events less than one tenth of a second apart, only those mouse events greater than one-tenth of a second were considered for hesitant-style pause. Another related limitation was the least count feature of recordable time stamps. Any mouse event less than 15 ms apart was recorded with a same time stamp as the previous event, resulting in a zero time interval scenario, whenever user performed a quick jerky movement. This was a system level limitation and only non-zero intervals were considered.

9.4 Mouse Interactivity in Multimodal Measures

A pointing device such as a mouse may not be used continuously throughout typical user sessions. However this is also the case with other device/channel/modalities. It is because of this that multimodal measures are grouped together for better cognitive load detection. But mouse interactivity does have its strong points. First of all, it is non-intrusive and actually an embedded part of the whole task scenario. Capturing mouse activity does not require any additional hardware and that is very conveniently non-intrusive as compared to other type of scenarios that eg maybe recording eye activity or EEG signals. Also that mouse activity signals are very sensitive to any voluntary or involuntary hand movement, even if the control buttons are not being pressed or used. This spatial dimension of mouse activity is good at analysing hand movement patterns in different load conditions.

9.5 Summary

These experiments showed how the participants used the mouse under different levels of cognitive load. It can be concluded that pause/break in mouse activity is a feature that holds significant promise for detecting high cognitive load on working memory. Contemplation-style pause has the potential to be treated as a separate feature; however, more investigation needs to be carried out with regards to its being a direct indicator of cognitive load variation. Hesitation-style pause features show significant differences and, coupled (or interpreted) with trajectory based features, can be used to predict variations in cognitive load. Finally, the number of lines, where the participants moved the mouse horizontally and vertically, was significantly lower under a high cognitive load than under a low cognitive load.

References

1. F. Chen, N. Ruiz, E. Choi, J. Epps, M.A. Khawaja, R. Taib, B. Yin, Y. Wang, Multimodal behavior and interaction as indicators of cognitive load. ACM Trans. Interact. Intell. Syst. **2**(4), 22:1–22:36 (2012)
2. Z. Jorgensen, T. Yu, On mouse dynamics as a behavioral biometric for authentication, in *Proceedings of the 6th ACM Symposium on Information, Computer and Communications Security*, New York, NY, USA, 2011, pp. 476–482
3. S. Arshad, Y. Wang, F. Chen, Analysing mouse activity for cognitive load detection, in *Proceedings of the 25th Australian Computer-Human Interaction Conference: Augmentation, Application, Innovation, Collaboration*, 2013, pp. 115–118
4. R. Brunken, J.L. Plass, D. Leutner, Direct measurement of cognitive load in multimedia learning. Educ. Psychol. **38**(1), 53–61 (2003)
5. N. Ruiz, R. Taib, F. Chen, Examining the redundancy of multimodal input, in *Proceedings of the Annual Conference of the Australian Computer-Human Interaction Special Interest Group (OzCHI 2006)*, Sydney, Australia, 2006
6. D.A. Schulz, Mouse curve biometrics, in *Biometric Consortium Conference, 2006 Biometrics Symposium: Special Session on Research at the*, 2006, pp. 1–6
7. B. Yin, F. Chen, Towards automatic cognitive load measurement from speech analysis, in *Human-Computer Interaction, Interaction Design and Usability*, ed. by J. Jacko (Springer, Berlin, 2007), pp. 1011–1020
8. A. Khawaji, F. Chen, J. Zhou, N. Marcus, Trust and cognitive load in the text-chat environment: The role of mouse movement, in *Proceedings of the 26th Australian Computer-Human Interaction Conference on Designing Futures: The Future of Design*, New York, NY, USA, 2014, pp. 324–327
9. L.E. Scissors, A.J. Gill, K. Geraghty, D. Gergle, In CMC we trust: The role of similarity, in *Proceedings of the 27th international conference on Human factors in computing systems*, 2009, pp. 527–536
10. N. Bos, J. Olson, D. Gergle, G. Olson, Z. Wright, Effects of four computer-mediated communications channels on trust development, in *Proceedings of the SIGCHI Conference on Human Factors in Computing Systems*, New York, NY, USA, 2002, pp. 135–140
11. NASA. Nasa Task Load Index, Version 1.0, 1986
12. *Math Open Reference*. [Online]. Available: http://www.mathopenref.com/tocs/coordpointstoc.html. Accessed 15 Sept 2015

Part IV
Multimodal Measures and Affecting Factors

Chapter 10
Multimodal Measures and Data Fusion

Multimodal interaction continues to be the preferred mode in the world of interface design [1]. The rapidly growing world of digital devices and online learning has demanded attention be paid to multimodal interaction challenges. For example, to learn efficiently (or to engage in error-free interaction), a typical user interface must intelligently handle multimodal measures and also fuse data from several sources in order to provide a meaningful interpretation. This chapter focuses on:

- Setting up a model for the multimodal cognitive load assessment;
- A user study of basketball skills training is presented to show the applicability of the model in multimodal data processing;
- Performance (accuracy) measures, physiological signals (GSR) and subjective ratings are collected to establish a ground truth for cognitive load and task difficulty;
- Pen input, speech and GSR are fused using the AdaBoost boosting algorithm in the multimodal fusion.

10.1 Multimodal Measurement of Cognitive Load

Given previous successes in finding features from various modalities such as pen and speech input that allow us to differentiate cognitive load levels the next step that suggests itself is to apply a multimodal index of load that combines output from different sources. Correlations between single-modality indices offer a way in which to introduce redundancy and robustness to a multimodal index of cognitive load. Dual-modality indices working together in a complementary fashion, such as speech signal based classification or degree of degeneration of pen input are likely to align quite well, reinforcing each other. However, there are a number of aspects that need to be considered in the development of a multimodal index of load, for

F. Chen et al., *Robust Multimodal Cognitive Load Measurement*, Human–Computer Interaction Series, DOI 10.1007/978-3-319-31700-7_10

example, whether early or late fusion approaches are used. At an abstract level, multimodal indices can be derived in four ways [2]:

- Combining component features within each modality eg combining within pen-input features such as stroke frequency, Mahalanobis distance (MDIST) [3], [4] or altitude span;
- Combining component features across modalities eg combining stroke frequency (from a pen) with use of singular pronouns (linguistic);
- Combining index results between modalities, eg between pen-only assessment vs. speech signal-only assessment;
- Using a combination of any of the three methods above.

10.2 An Abstract Model for Multimodal Assessment

Figure 10.1 depicts a high level functional model of a proposed Cognitive Load Measurement (CLM) system [1]. The abstract system model embodies four high level processes: pre-processing and data cleaning, feature extraction, load assessment and index fusion. The great advantage of multimodal behavioral indices of cognitive load is that they are derived from activity already undertaken as part of the task, and thus can be collected implicitly, or "passively" [5]. The raw modality input sources are first and foremost intended for purposes other than cognitive load measurement, specifically to do with the domain application. For example, the data may be used for semantic interpretation or rendering (eg in the case of command and control speech or interactive pen gestures). The data may therefore need to be duplicated and diverted – with the original stream sent to the recognizers, and a secondary stream sent to the Cognitive Load Measurement engine. In Fig. 10.1, speech input data is first captured through a close-talk microphone. This generates two kinds of data, speech signal data (eg a wav file) and text (through a speech to text engine). Likewise, pen input data is collected as trajectory tuples, including pressure, pen orientation and other information transmitted directly from the device drivers, alongside system timestamps.

Data pre-processing and data cleaning refers to any reformatting, restructuring of the input data, or removal of unnecessary information, for example, any outliers or segments that are too short for geometric and temporal analysis; words not recognized in the text, as well as words that are not used in the analysis. Input streams from other modalities will follow the same processes. Similarly, a number of other non-behavioral indices will also undergo pre-processing as needed; these include indices that may also be used in the process, such as galvanic skin response, or other body-based data, such as posture, movement or temperature. Environmental and other external context information may also be provided to the CLM system for enhanced performance at this point.

The second stage involves streaming the individual modal input into their respective feature extraction components. The same data may be used for multiple

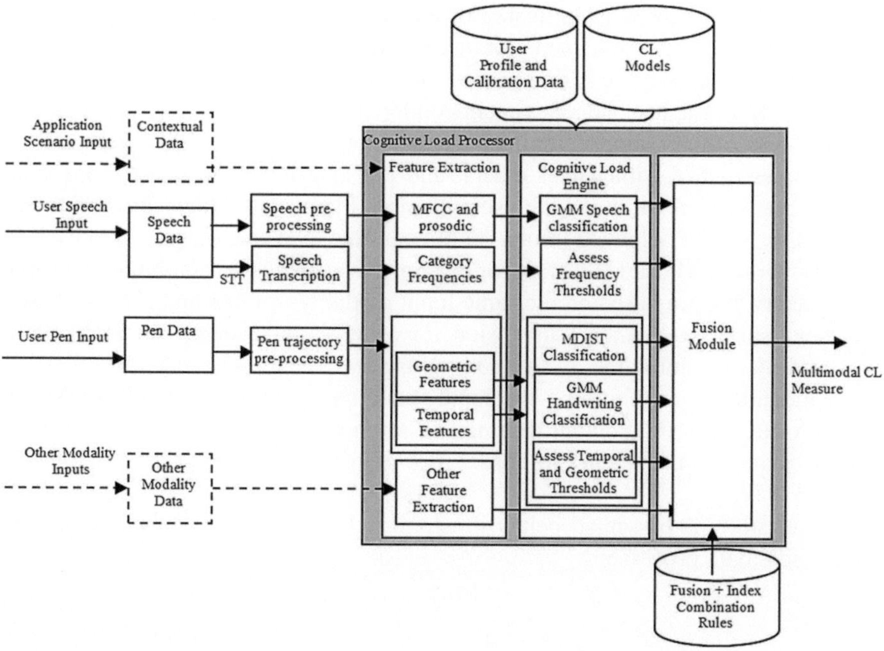

Fig. 10.1 High level functional model of a multiple modality CLM system [1]

feature extraction components, while other extraction components may not be activated, depending on domain-specific contextual information gathered from the active applications and workflow diagrams established a-priori. This will allow the feature extraction engine to choose the most appropriate modules to activate for each incoming input stream. For example, if the incoming speech is sourced from a phone call or radio conversation, the feature extraction component will activate both MFCC and prosodic feature extraction as well as the linguistic category extraction components, since both can provide meaningful measures on this kind of data. On the other hand, if the incoming speech is sourced from command and control input, only MFCC and prosodic feature extraction will be activated, as the linguistic categories cannot provide any meaningful cognitive load measurement information on short, closed vocabulary, single word speech.

The third stage involves the decision-making aspect of the process, where thresholds are invoked and the appropriate models for each modality are selected from the database from which to carry out the classification. For example, for the speech signal based cognitive load measurement, different models are required for single word cognitive load classification versus continuous speech classification. Likewise, different MDIST models exist for each shape, and also for each user. Any calibration data that is needed for classification or for comparison purposes is also accessed at this point.

The final stage involves the fusion of indices resolved from the previous stage. The assessment results obtained from each modality can also convey confidence information to support the fusion process. The fusion engine accesses information regarding the modality load assessment combination rules in each specific context, eg whether the time-windows for the collected inputs are compatible; which indices are complementary with which others; and the appropriate weightings for each index, given the scenario and the user situation. Figure 10.1 shows how mid- and late- fusion may be achieved from a set of cognitive load assessments from each of the sub-features. Mid-level fusion, for example, is achieved by combining multiple assessments that are based on the same input modality, for example, speech based and linguistic assessments. Late fusion for a multimodal index can be likewise achieved by combining the results from all the features individually (regardless of input modality), or combining the input modality subgroup from the mid-level fusion results. The final output from the CLM engine can then be passed onto the output generation system in order to implement appropriate adaptation strategies [1].

A user study is presented below illustrating the applicability of this model to multimodal data processing.

10.3 Basketball Skills Training

In order to illustrate how a multimodal cognitive load measurement system could work, a lab-based study is presented in which cognitive load and complexity were manipulated, and multiple behavioral modalities were recorded. The objective is to assess how well individual and combined modalities can reflect levels of cognitive load, and provide a concrete application for the multimodal cognitive load measurement model. While the task is different to the safety-critical, data-laden, and high-intensity applications as discussed in other chapters of the book, it is a richly multimodal data set that helps to provide an example application of the model presented in the previous section. Elite athletes at the Australian Institute of Sport (AIS) are required to complete cognitive skills training using a targeted sports-specific software application called AISReact [6]. While aiming at ever faster situation analysis and decision making through the construction better mental schemas, it is desirable to precisely determine onsets of very high cognitive load in order to adapt the training rate to each individual athlete. In this experiment, we modified the software to accept pen based interaction, and added the modalities of speech and eye-activity. In addition, performance (accuracy) measures, GSR and subjective ratings were also collected to establish a ground truth for cognitive load and task difficulty [1]. The setup is shown in Fig. 10.2.

Twelve male recreational basketball players, aged 19–36, each with more than 2 years' experience (average of 9.4 years) volunteered to complete the study. The task consisted of a 10 s video basketball clip played on a tablet monitor, which was then frozen and replaced with a blank court schematic. The clips involved 10 players

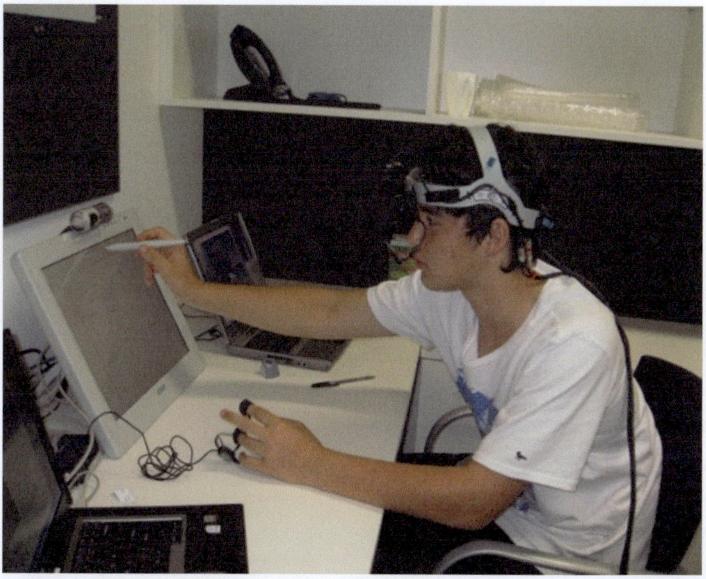

Fig. 10.2 Physical set-up of a user completing a task using a digital pen and with GSR attached

and the participants had to remember the locations and roles of some players in three task difficulty levels (remember 3 players for Low level, 6 for Medium, and all 10 for High). Each level consisted of 6 distinct clips. The clips were filmed from above and cover half the court, with all plays moving from the bottom of the screen towards the top, where the basketball hoop was located, as seen in Fig. 10.3.

The participants used specific pen marks to identify the remembered player positions on the tablet monitor: attackers were denoted by crosses, defenders by circles and the ball carrier by a circle with a dot in the middle, as illustrated in Fig. 10.3. Participants were also instructed to think aloud through their answers, and these utterances were captured using a close-talk microphone.

10.4 Subjective Ratings and Performance Results

Subjective ratings were collected using a Likert 9-point scale, where 1 was minimal effort and 9 was extreme effort. The task complexity levels induced extreme levels of load as reflected in the subjective ratings, increasing significantly as cognitive load increased, with mean averages of 3.2 (SD = 1.34), 5.5 (SD = 1.62) and 7.6 (SD = 1.23) for the Low, Medium and High load tasks respectively in Fig. 10.4. Due to the non-parametric dataset, this was verified using Friedman's $\chi 2$ test ($\chi 2$ (12,2) = 25.53, $p < 0.001$), where Low, Medium and High were ranked 1.00, 2.04 and 2.96 respectively.

Fig. 10.3 The clips example in the study: (**a**) Last frame of video clip before freeze, (**b**) Blank court image with player markings

As expected, the participants' performance decreased significantly from Low load to High load. Scores were given for each mark whose centroid was placed within a radius of 8 % screen distance (in pixels) from the correct player position, as recommended by basketball experts at the Australian Institute of Sport, who also annotated the correct player positions on the schematic. The mean score for the Low, Med and High load tasks 83.5 % ($SD = 11.63$), 77.7 % ($SD = 12.26$) and 68.1 % ($SD = 15.14$). The decrease was verified through a repeated-measures ANOVA test ($F(2,22) = 4.84$, $p = .018$). Subsequent planned contrasts show a significant linear ($F(1,11) = 5.59$, $p = 0.04$, $r = 0.46$) to the 0.05 level, with a medium effect size. This is evident in Fig. 10.4 also, where the performance

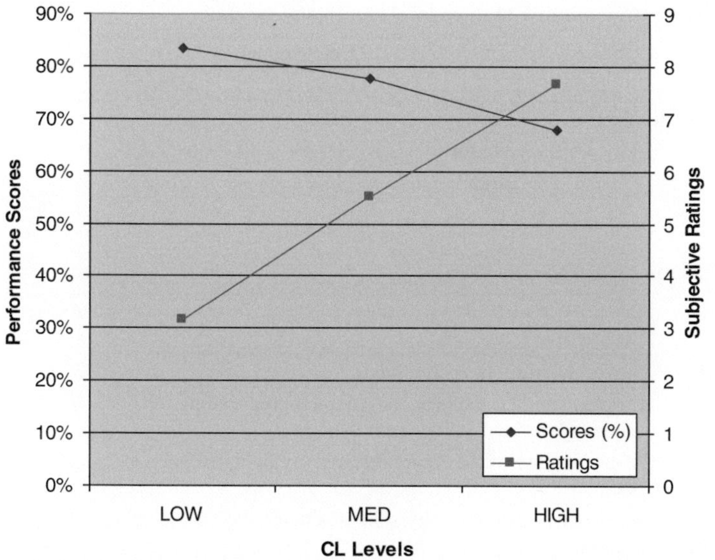

Fig. 10.4 Performance scores and subjective ratings [7]

decreases gradually between Low and Medium load levels and then more steeply from Medium to High levels.

Overall, participants' performance decreased significantly, while their subjective ratings of load increased significantly, from Low load to High load, validating that the responses elicited by these tasks are affected by extreme levels of cognitive load.

10.5 Individual Modalities

In this section, the capacity of individual modalities is analyzed to classify load levels. In addition to speech and pen input, GSR is presented although it is a physiological measure, because the relative potential of these three modalities is compared in the next section. In this analysis, a cognitive load estimate was made at the end of the task, ie after around 1–5 s.

Speech data was analyzed for all 12 subjects [7], and the results use the average of the two evaluation folds, classifying into three pre-designed load levels. As shown in Table 10.1, Low load achieved 100 % accuracy, and High load 82 % of testing samples. Interestingly however, testing samples from the Medium load level were mostly misclassified into either the Low or High load, suggesting that no distinct pattern was captured. It is suspected that participants with subtly varied basketball skills and load capacity may have experienced slightly lower or higher loads in this level. The average accuracy for the 3 levels was 62.7 %.

Table 10.1 Confusion matrix of three-level speech classification

		Classified as		
		Low (%)	Medium (%)	High (%)
Testing samples from	**Low**	100	0	0
	Medium	40	6	54
	High	15	3	82

Table 10.2 Pen-input trajectory features

Geometric feature	Description	Accuracy on test samples (%)
Duration	Stroke duration, in milliseconds	**32.6**
Length	Cumulative distance between sampled points along the trajectory	**40.7**
Mean velocity	Mean velocity of the stroke trajectory, calculated point to point	**30.7**
Mean acceleration	Mean acceleration of the stroke trajectory, calculated point to point	**37.0**
Area	The area in pixels taken by the circle shape, enclosed by the trajectory	**36.3**
First-Last	Distance between the first and last points of the trajectory	**33.3**
Overlap ratio	The ratio of the overlapping distance between the first and last points of the trajectory to the total size of the shape.	**37.4**

Unfortunately, due to corrupt collected signals from some of the GSR input sensor, and data losses caused by unexpected crashes in the software, only 9 subjects have complete data for the purpose of fusion, and hence this subset will be exclusively used for the remainder of this case study. For these 9 subjects only, the average speech classification accuracy drops slightly, to 61.8 %.

Pen input was analyzed through a set of simple, objective, features based on circling shapes drawn by the participants. Table 10.2 summarizes the features and their individual accuracy at classifying load levels for the 9 subjects. The results range from 31 to 41 % for the 3-level classification, ie in some cases not always outperforming chance classification (which would result in an accuracy of 33.3 %).

Although galvanic skin response is not a behavioral measure of cognitive load but a physiological one (ie it is not a voluntary reaction but a function of the autonomic nervous system), it was used as a ground truth measurement for the study. Measured in micro-Siemens (μS), the signal was simply analyzed using an average measurement over the task period, yielding a classification accuracy of 64.4 % over 3 load levels, across all 9 subjects, using a leave-one out evaluation scheme.

10.6 Multimodal Fusion

In this section, the above features extracted from speech, pen input and GSR are fused using the AdaBoost boosting algorithm. Boosting [8], [9] is a general ensemble learning algorithm, which creates an accurate strong classifier H by iteratively combining a number T of moderately inaccurate weak classifiers h_t. By definition, a strong classifier has high classification accuracy on the data set, while a weak classifier's accuracy is just above that of a random guess. The final strong classifier can be defined as:

$$H(x) = \begin{cases} 1 & \text{if} \sum_{t=1}^{T} \alpha_t h_t(x) \geq 0 \\ -1 & \text{otherwise} \end{cases}$$

where α_t is a weight coefficient. In simple cases, each weak classifier is attached to a feature, so the process of combining weak classifiers in Boosting is equivalent to a feature fusion process.

This study used AdaBoost [10], [11], an adaptive version of boosting. Sample weights are all initially set equal, then refined iteratively during a training process. In order to select those features that are most discriminative of a given problem, in each iteration AdaBoost selects a new weak classifier h_t with the minimal weighted classification error with respect to the training sample weight distribution, which means the newly selected weak classifier can guarantee the more important samples (samples with higher weights) are classified correctly. Then the weights of incorrectly classified samples are increased, so in the next iteration, AdaBoost can focus on these incorrectly classified samples.

Table 10.3 details the weights obtained for speech and the pen features. The average classification accuracy when fusing all these features is 64.1 % on the testing samples, for the 3 load levels across all 9 subjects. It should be noted that this represents a small improvement over the speech-only accuracy in the previous section. Cognitive load classification of free-form pen features is a challenging task (this is seen also in Table 10.2), however the prospects for handwriting features are considerably more positive.

Similarly, Table 10.4 details the weights obtained when fusing speech, pen features and also GSR. The average classification accuracy is then 77.8 % on the testing samples, for the 3 load levels across all 9 subjects.

Adding the GSR feature provides a significant improvement, supporting the benefits of feature fusion for workload detection. This case study is proposed as one implementation example of the model, however the results indicate that other

Table 10.3 AdaBoost weights for speech and pen input features

	Speech	Duration	Length	Velocity	Acceleration	Area	First-Last	Overlap
Weights	0.686	0.150	0.053	0.000	0.051	0.059	0.000	0.002

Table 10.4 AdaBoost weights for speech, pen input features and GSR

	Speech	Duration	Length	Velocity	Acceleration	Area	First-Last	Overlap	GSR
Weights	0.478	0.176	0.055	0.05	0.041	0.053	0.011	0.000	0.181

behavioral features, yet to be explored, may be able to provide further multimodal cognitive load measurement accuracy.

10.7 Summary

Previous chapters focused on finding features from various single modalities such as GSR, pen and speech input that allow us to differentiate cognitive load levels. Correlations between single-modality indices offer a way in which to introduce redundancy and robustness to a multimodal index of cognitive load. Multimodality indices working together in a complementary fashion may reinforce each other and improve the performance of cognitive load indexing. This chapter presented a model for the multimodal cognitive load assessment that combines output from different sources. A user study was presented to show the applicability of the model in multimodal data processing. The features extracted from speech, pen input and GSR were fused using the AdaBoost boosting algorithm.

References

1. F. Chen, N. Ruiz, E. Choi, J. Epps, M.A. Khawaja, R. Taib, B. Yin, Y. Wang, Multimodal behavior and interaction as indicators of cognitive load. ACM Trans. Interact. Intell. Syst. **2**(4), 22:1–22:36 (2012)
2. N. Ruiz, Cognitive load measurement in multimodal interfaces. PhD Thesis, University of New South Wales, Sydney, Australia, 2011
3. N. Ruiz, R. Taib, F. Chen, Freeform pen-input as evidence of cognitive load and expertise, in *Proceedings of the International Conference on Multimodal Interfaces (ICMI 2011)*, 2011
4. D. Rubine, Specifying gestures by example, in *Proceedings 18th annual conference on Computer graphics and interactive techniques (SIGGRAPH 1991)*, New York, USA, 1991
5. T.O. Zander, C. Kothe, Towards passive brain-computer interfaces: Applying brain-computer interface technology to human-machine systems in general. J. Neural Eng. **8**(2), 025005 (2011)
6. C. Mackintosh, *AIS React Software v.6.6*. Australian Institute of Sport, 2010
7. N. Ruiz, G. Liu, B. Yin, D. Farrow, F. Chen, Teaching athletes cognitive skills: Detecting load in speech input, in *Proceedings of the 24th BCS Conference on Human Computer Interaction (HCI 2010)*, Dundee, Scotland, 2010
8. Y. Freund, Boosting a weak learning algorithm by majority. Inf. Comput. **121**(2), 256–285 (1995)
9. R.E. Schapire, The strength of weak learnability. Mach. Learn. **5**(2), 197–227 (1990)
10. R.E. Schapire, Y. Freund, P. Bartlett, W.S. Lee, Boosting the margin: A new explanation for the effectiveness of voting methods. Ann. Stat. **26**(5), 1651–1686 (1998)
11. Y. Freund, R.E. Schapire, A decision-theoretic generalization of on-line learning and an application to boosting. J. Comput. Syst. Sci. **55**(1), 119–139 (1997)

Chapter 11
Emotion and Cognitive Load

In practice, various confounding factors unrelated to workload, including changes of luminance condition and emotional arousal may degrade workload measures such as commonly used mean pupil diameter [1, 2]. The highlights of this chapter include:

- Using pupillary response and GSR as a cognitive load measure under the influence of luminance and emotional arousal confounding factors. A video-based eye tracker is used to record pupillary response during arithmetic tasks under luminance and emotional changes;
- Machine learning based feature selection and classification techniques are presented to robustly index cognitive workload based on pupillary response and GSR even with the influence of noisy factors unrelated to workload.

11.1 Emotional Arousal and Physiological Response

One of the common elements influencing human physiological response is emotional arousal. Stanners et al. examined two psychological interpretations of pupillary response, namely emotional arousal and cognitive workload and found that cognitive workload took priority over emotional factors in affecting pupillary response [3]. Bradley et al. [4] conducted a picture viewing experiment which utilized a set of selected pictures from the International Affective Picture System (IAPS) [5]. They found that under controlled luminance condition, pupillary changes became larger when participants viewed emotionally arousing pictures, regardless of whether these pictures were pleasant or unpleasant. Partala and Surakka [6] examined pupil size variation during and after auditory emotional stimulation. The experimental results showed that compared with the process involving neutral stimuli, pupil size was significantly larger during the process involving emotionally positive or negative stimuli.

F. Chen et al., *Robust Multimodal Cognitive Load Measurement*, Human–Computer Interaction Series, DOI 10.1007/978-3-319-31700-7_11

As reviewed in the previous chapters, GSR has also been investigated as a tool for the measurement of mental status involving cognitive workload and other factors. Shi et al. [7] evaluated the correlation between GSR data and cognitive workload in a multimodal user interface. Mean GSR and accumulated GSR have been used to measure cognitive workload during traffic control management tasks. In addition, GSR based morphological features including amplitude, recovery time and latency have been employed for stress measurement during cognitive processes [8]. It was reported that GSR can be affected by emotion based arousal as well. In Bradley et al.'s experiment [4], skin conductance increased when participants viewed neutral pictures compared to when they viewed unpleasant or pleasant pictures.

11.2 Cognitive Load Measurement with Emotional Arousal

In this section, pupillary response and GSR based workload measurements under changes of both luminance condition and emotion arousal are investigated [1, 9].

11.2.1 Task Design

In this experiment, 12 24-to-35-year-old male participants perform arithmetic tasks under changes to the luminance condition of a visual display and emotional arousal [9].

A remote eye tracker was used to record pupillary response data and a GSR sensor (ProComp Infiniti of Thought Technology Ltd) was used to record skin conductance data. The experiment setup is demonstrated in Fig. 11.1.

The experiment was composed of three parts. In the first part, the arithmetic task was carried out with a blank background (black screen). In the second and third parts, the arithmetic task was carried out with pleasant and unpleasant background images shown on the screen.

As with the arithmetic task in Sect. 5.2, during each arithmetic task, the subjects were asked to sum up four different numbers and then choose the correct answer via a mouse click. At the beginning of a task, the same "X" displaying period appeared for 3 s before the first number was displayed. The arithmetic tasks had the same four task difficulty levels as those in Sect. 5.2. The difference between the experiment in Sect. 5.2 and this experiment comes from the setting of the background condition. In the first part of the experiment in this chapter, a blank background was displayed on the screen. In the second and the third parts, pleasant and unpleasant background images were displayed during the tasks. Eight pleasant images (mean valence/arousal = 7.1, 5.7) and eight unpleasant images (mean valence/arousal = 2.8, 4.8) were selected from the IAPS database. The mean luminance (Y value) of the images ranged from 53 to 174.

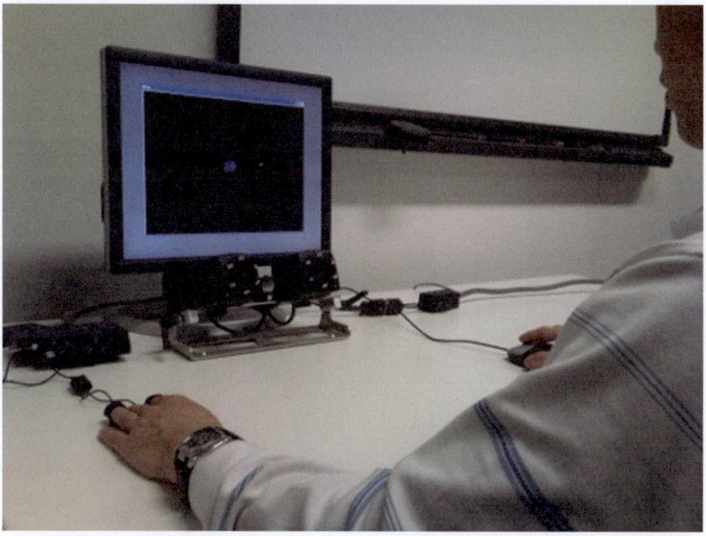

Fig. 11.1 Experiment setup

At the beginning and the end of the experiment, each subject was given a 1-min rest with a black screen. Eight arithmetic tasks were assigned in random order in each experiment part (two for each difficulty level). The experiment lasts about 15 min for each subject.

11.2.2 Pupillary Response Based Measurement

The average pupil diameters under different task difficulty levels and background conditions are presented in Fig. 11.2. It can be seen that with the blank background, pupil diameter increased when the task difficulty level rises. However this trend was not observed for the pupil diameters in the pleasant and unpleasant background image conditions, presumably because the pupillary response was affected by the changes of both luminance condition and emotional arousal. It can be seen in the figure that the pupil diameter under the highest task difficulty with emotional background images was smaller than that under the lowest task difficulty with the blank background. The phenomenon is again consistent with the observation of previous empirical research that the effect of luminance conditions overshadows changes in pupil diameter induced by cognitive demands and emotional arousal. Thus we can conclude that average pupil size or dilation may not be an effective tool to measure cognitive workload when under the influence of noisy factors including luminance variation and emotional arousal [1, 2].

The same mean-difference feature as investigated in Sect. 4.4 was used to measure cognitive load based on the pupillary response data. The distribution of

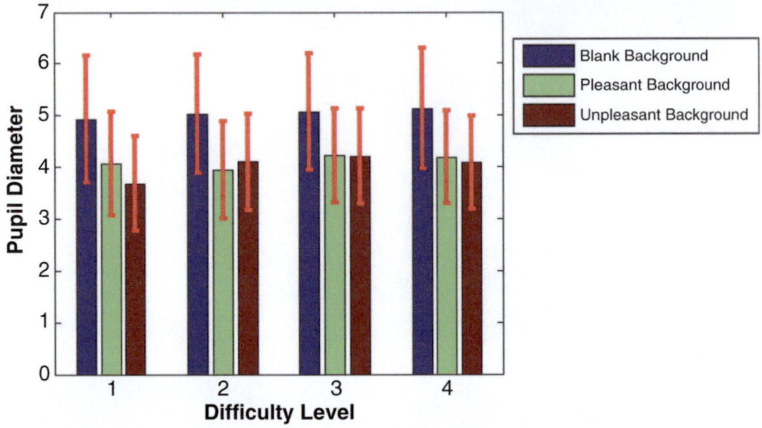

Fig. 11.2 Pupil diameter under different task difficulty levels and background conditions

the feature values corresponding to different task difficulty levels is shown in Fig. 11.3. From this we can see that even with the change of background luminance and emotion arousal, the pupillary response based feature value increases as the task difficulty level increases, and statistical significance ($F = 3.59$, $p < 0.05$) is found in the corresponding ANOVA test.

Figure 11.4 shows the distribution of mean-difference feature values under different background conditions (blank, pleasant and unpleasant background images). The mean-difference feature does not exhibit significant difference under different conditions of emotional arousal ($F = 1.09$, $p = 0.35$ in corresponding ANOVA test). These results indicate that compared with cognitive load, the pupillary response based feature was less sensitive to emotional arousal.

11.2.3 Skin Response Based Measurement

The skin response of each subject was also recorded during the experiment. This section investigates the GSR based measurement of cognitive load under the influence of emotional arousal.

Figure 11.5 shows the distribution of average GSR values during the task period under different task difficulty levels and background conditions. It can be seen that the average GSR value cannot effectively distinguish different task difficulty levels ($F = 0.01$, $p = 0.99$ in the corresponding ANOVA test). Similarly, the mean-difference feature was used to attempt to characterize cognitive workload based on GSR data. Figure 11.6 shows the distribution of GSR based feature values under different task difficulty levels. While it appears that overall the feature value tends to decrease with increasing task difficulty level, statistical significance was not found in the corresponding ANOVA test ($F = 1.83$, $p = 0.16$).

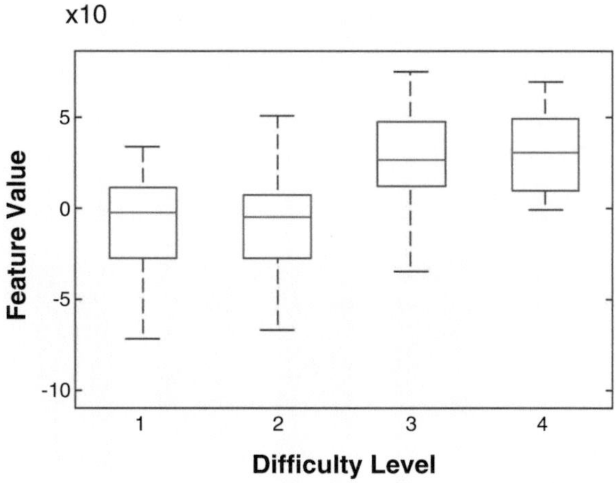

Fig. 11.3 Box plot of pupillary response based feature values corresponding to different task difficulty levels

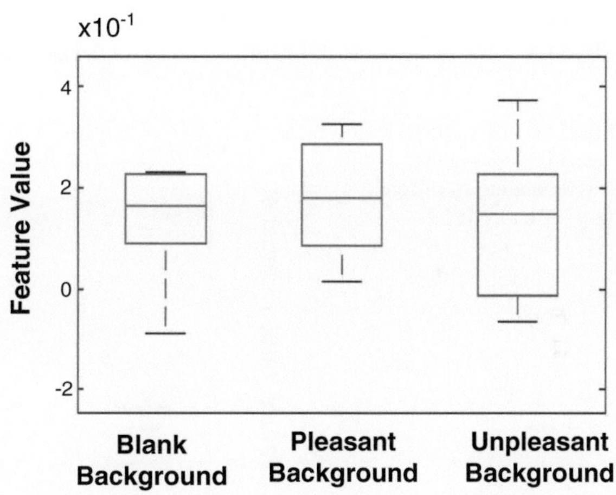

Fig. 11.4 Box plot of pupillary response based feature values corresponding to different background conditions

Figure 11.7 also shows the distribution of mean-difference feature values under different background conditions. The mean-difference feature demonstrates significant difference under different conditions of emotional arousal ($F = 3.78$, $p < 0.05$ in the corresponding ANOVA test). Hence the statistical results indicate that compared with cognitive load, the GSR based feature is more sensitive to emotional arousal in the experiment.

11.3 Cognitive Load Classification with Emotional Arousal

Cognitive load classifications were performed under different conditions. Boosting algorithms presented in Sect. 4.4 were used for cognitive load classifications.

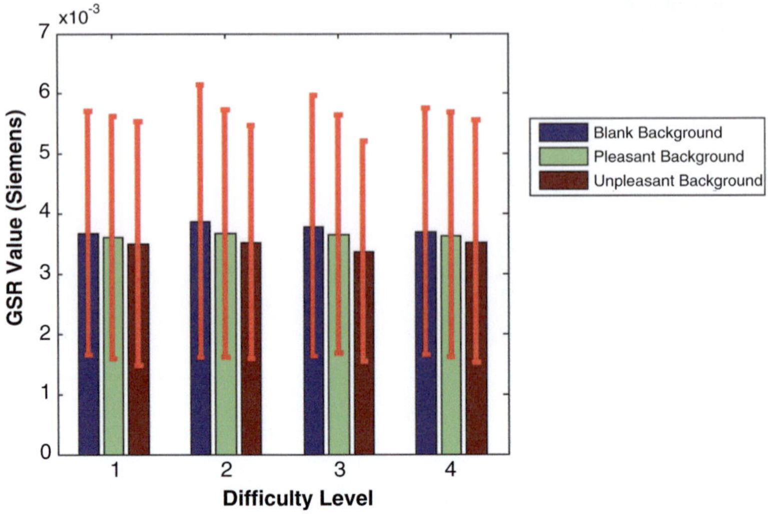

Fig. 11.5 GSR value under different task difficulty levels and background conditions

Fig. 11.6 Box plot of GSR based feature values corresponding to different task difficulty levels

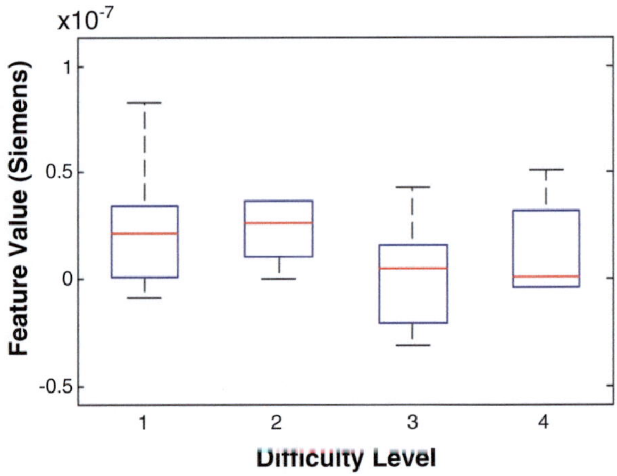

11.3.1 Cognitive Load Classification Based on Pupillary Response

Table 11.1 shows the accuracies of two-class and four-class classification based on pupillary response data from the experiment. It shows that Boosting can achieve about 80 % testing accuracy for two-class classification and about 43 % for four-class classification.

Figures 11.8 and 11.9 demonstrate the distributions of selected Haar-like features in terms of feature temporal size *r* and centre location *t* respectively for four-

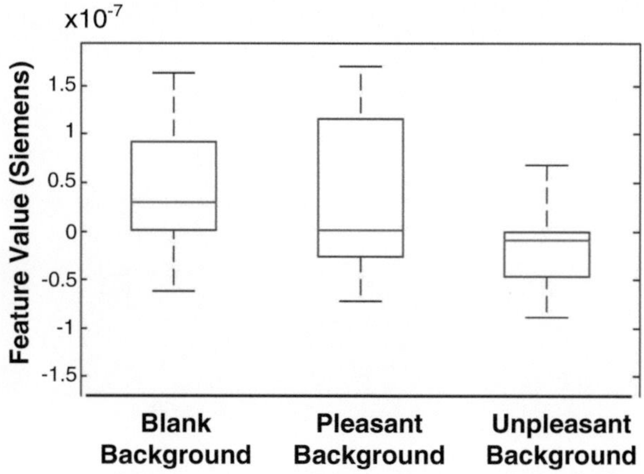

Fig. 11.7 Box plot of GSR based feature values corresponding to different background conditions

Table 11.1 Workload classification accuracy based on pupillary response

	Two-class (%)	Four-class (%)
Training accuracy	90.8	61.2
Testing accuracy	79.2	43.2

class classification. It was seen that the Boosting algorithm tended to select features with large temporal size and features located in the middle of task period, which indicates that those features are relatively discriminatory for workload classification.

11.3.2 Cognitive Load Classification Based on GSR

GSR data from the experiment were also tested by the Boosting algorithm with Haar-like features as presented in Sect. 4.4 for cognitive load classifications. Table 11.2 shows accuracy of two-class and four-class classification. It shows that neither two-class nor four-class classification accuracy was as good as those based on pupillary response, which indicates that pupillary response is more effective than GSR for workload classification. This observation is consistent with later experimental results based on the fusion of pupillary response and GSR data.

The distributions of selected Haar-like features for four-class classification are presented in Figs. 11.10 and 11.11. Similar to the experimental statistics based on pupillary response, features located in the middle of the task period are more likely to be selected. However, for GSR based features, those with small temporal size are more likely to be selected, which is the opposite of the statistical results we saw based on pupillary response.

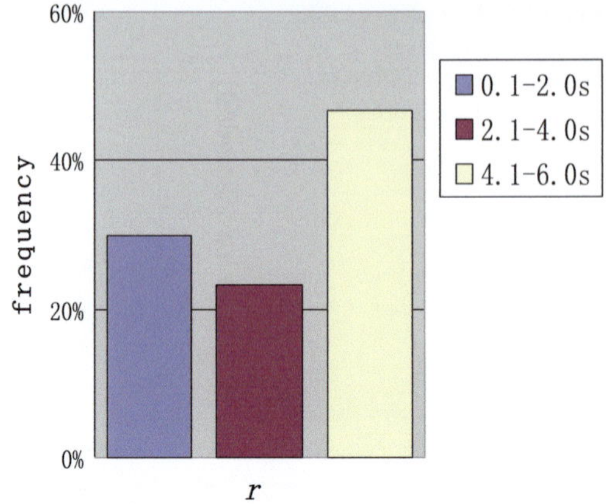

Fig. 11.8 Distribution of feature size for Haar-like features used in workload classification based on pupillary response

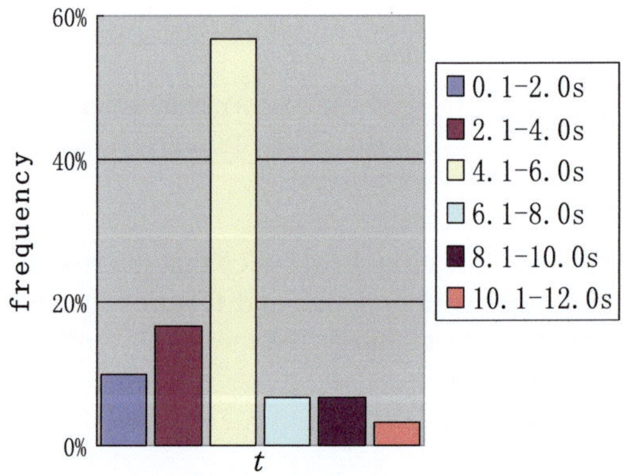

Fig. 11.9 Distribution of feature centre for Haar-like features used in workload classification based on pupillary response

Table 11.2 Workload classification accuracy based on GSR for experiment 2

	Two-class (%)	Four-class (%)
Training accuracy	79.3	47.0
Testing accuracy	67.3	36.9

11.3.3 Cognitive Load Classification Based on the Fusion

In this section we discuss pupillary response data and GSR data that were combined to attempt to classify cognitive load levels. Table 11.3 shows the accuracy of two-class and four-class classification from a fusion of pupillary response data and GSR data from the experiment. Data fusion was achieved by the Boosting algorithm. Haar-like features extracted from both pupillary response data and GSR data were combined for feature selection and workload classification during the classifier training process. The classification performance based on the fused data was slightly better than that based on the pupillary response data alone and much better than that based on the GSR data alone, since data fusion can provide richer feature space for classifier construction.

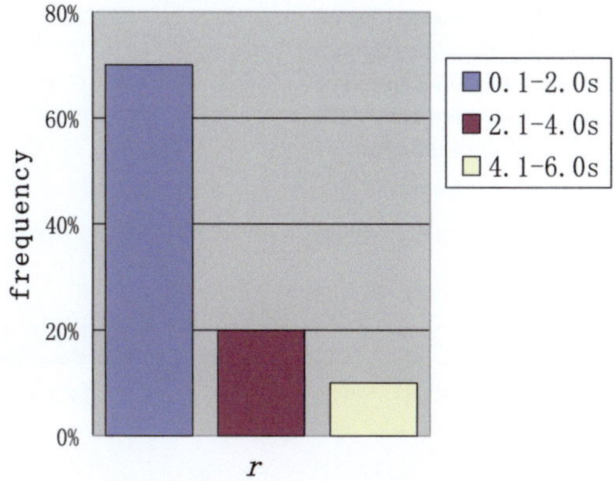

Fig. 11.10 Distribution of feature size for the Haar-like features used in workload classification based on GSR

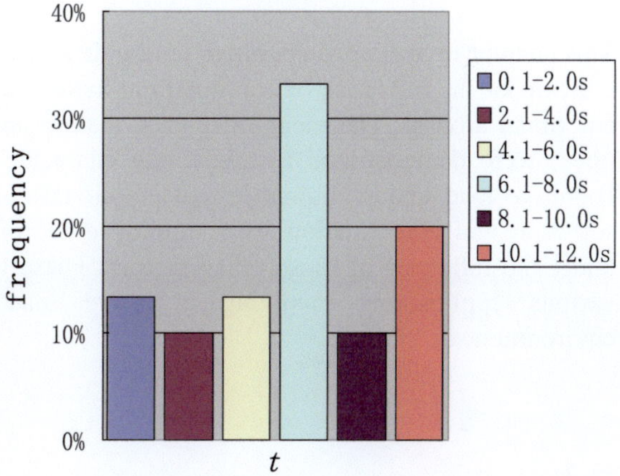

Fig. 11.11 Distribution of feature centre for the Haar-like features used in workload classification based on GSR

Table 11.3 Workload classification accuracy based on the fusion of pupillary response and GSR data

	Two-class (%)	Four-class (%)
Training accuracy	92.4	56.6
Testing accuracy	82.1	45.8

In Fig. 11.12, the proportion of pupillary response based features was compared against that of GSR based features used by the two-class and four-class workload classifiers. It was found that the features extracted from pupillary response data were dominant. This indicates again that pupillary response is more useful than GSR for cognitive workload classification in our experiment.

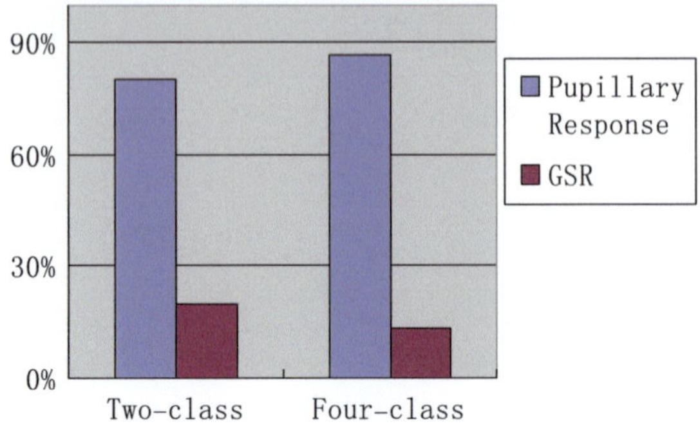

Fig. 11.12 Proportions of pupillary response based features and GSR based features selected for workload classification

11.4 Summary

This chapter investigated cognitive load indexing based on pupillary response and GSR under the influence of confounding factors such as luminance condition and emotional arousal. The mean-difference feature and its extension (Haar-like features) were demonstrated to effectively characterize physiological responses of cognitive load under luminance and emotional changes. Boosting based feature selection and classification were employed to robustly classify workload even under the influence of those noisy factors. These techniques could be applied to various applications involving cognitive load evaluation under complex environments.

References

1. W. Wang, Z. Li, Y. Wang, F. Chen, Indexing cognitive workload based on pupillary response under luminance and emotional changes, in *Proceedings of the 2013 International Conference on Intelligent User Interfaces*, 2013, pp. 247–256
2. J. Xu, Y. Wang, F. Chen, E. Choi, Pupillary response based cognitive workload measurement under luminance changes, in *Human-Computer Interaction – INTERACT 2011*, ed. by P. Campos, N. Graham, J. Jorge, N. Nunes, P. Palanque, M. Winckler (Springer, Berlin/Heidelberg, 2011), pp. 178–185
3. R.F. Stanners, M. Coulter, A.W. Sweet, P. Murphy, The pupillary response as an indicator of arousal and cognition. Motiv. Emot. **3**(4), 319–340 (1979)
4. M.M. Bradley, L. Miccoli, M.A. Escrig, P.J. Lang, The pupil as a measure of emotional arousal and autonomic activation. Psychophysiology **45**(4), 602–607 (2008)
5. M.M.B.P.J. Lang, *International Affective Picture System (IAPS): Affective Ratings of Pictures and Instruction Manual*. Technical Report Rep. No. A-8 (University of Florida, Florida, 2005)
6. T. Partala, V. Surakka, Pupil zize variation as an indication of affective processing. Int. J. Hum. Comput. Stud. – Appl. Affect. Comput. Hum. Comput. Interact. **59**(1–2), 185–198 (2003)
7. Y. Shi, N. Ruiz, R. Taib, E. Choi, F. Chen, Galvanic Skin Response (GSR) as an index of cognitive load, in *CHI '07 Extended Abstracts on Human Factors in Computing Systems*, ed. by R. Mary Beth, pp. 2651–2656

8. C. Setz, B. Arnrich, J. Schumm, R. La Marca, G. Tröster, U. Ehlert, Discriminating stress from cognitive load using a wearable EDA device. IEEE Trans. Inf. Technol. Biomed. **14**(2), 410–417 (2010)
9. J. Xu, Y. Wang, F. Chen, H. Choi, G. Li, S. Chen, S. Hussain, Pupillary response based cognitive workload index under luminance and emotional changes, in *CHI '11 Extended Abstracts on Human Factors in Computing Systems*, New York, NY, USA, 2011, pp. 1627–1632

Chapter 12
Stress and Cognitive Load

Human physiological signals have been widely used to non-invasively measure cognitive load (CL) during task execution. A major challenge for CL detection is the presence of stress, which may affect physiological measurements in ways that confound reliable detection of CL. This chapter focuses on:

- Investigating the effect of stress on cognitive load measurement using GSR as a physiological index of CL;
- Utilizing feelings of lack of control, task failure and social-evaluation to induce stress;
- Features extracted from GSR signals based on peak detection exhibited consistent behaviour under both conditions, demonstrating the usefulness of the features as cognitive load index even when a person's stress level is fluctuating;
- Mean GSR values are shown to be significantly different between CL levels in the "no-stress" condition, but not when including the "stress" condition.

12.1 Stress and Galvanic Skin Response

Any given physiological signal used as an index for CL is likely to be affected by various additional inputs from the human body. Thus, a major task of cognitive load measurement via physiological means is demonstrating the diagnosticity and construct-validity of the CL index. One of the major contributors to change in human physiological systems is stress [1]. Stress has been shown to affect both the sympathetic and parasympathetic nervous systems and, in its more extreme states, results in large changes to physiological function that may well obscure the relationship between a physiological indicator and CL. Furthermore, stress may, in some circumstances, be a confounding factor for CL in that changes in CL may correlate with changes in stress levels. Construct-validity must be established

F. Chen et al., *Robust Multimodal Cognitive Load Measurement*, Human–Computer Interaction Series, DOI 10.1007/978-3-319-31700-7_12

before we can safely assert that changes in physiological indicators are the result of CL and not stress or other confounding factors [2].

Although definitions of stress vary, there is good consensus in the literature regarding conditions where it is likely to arise [3–5]. Failure at a task, together with feelings of lack of control, in situations where participants are evaluated by others is a widely used paradigm for stress induction. These conditions were operationalised in an experimental paradigm developed by Dedovic et al. called the "Montreal Imaging Stress Task" (MIST) [6]. The experiment presented here closely follows the MIST protocol with minor operational adjustments.

As reviewed in the previous chapters, GSR has been used successfully in the past to index CL. In an experiment involving traffic control management it was illustrated that the mean GSR of test subjects increases as the difficulty of cognitive tasks increases [7]. In addition, [8] analysed the time and frequency domains of recorded GSR signals and showed that CL imposed through arithmetic and reading tasks can be indexed by GSR of test subjects.

The relationship between GSR and stress has also been examined. In an experiment involving a driving task, Healey and Picard [9] were able to successfully classify different driving periods based on the stress levels of the driver. They extracted useful features from GSR signals recorded during the experiment based on peak detection and input these features into machine learning classification algorithms with positive results. [10] combined GSR with several other physiological signals to classify the stress states of test participants through the use of machine learning tools. Stress levels were induced by having subjects complete a "Paced Stroop Test", where the colour of a word that spells a different colour must be identified.

An interesting study on discriminating stress from cognitive load was carried out by Setz et al. [3]. However, they did not experimentally manipulate CL and only demonstrate the ability to differentiate between "stress" and "no-stress" conditions, where CL level was consistent between the two conditions. Nonetheless, the feature detection processes they outlined appear promising, and have been extended in the experiment presented here.

12.2 Cognitive Load Measurement Under Stress Conditions

12.2.1 Task Design

11 male students and employees (24–49 years old, ten right handed and one left-handed) took part in the experiment. Participants were offered one movie ticket and biscuits as recompense for their participation.

All participants had the voluntary nature of the experiment explained to them and then filled out a paper version of the Kessler K-10 Psychological Distress Scale

[11] to ascertain that they were unlikely to be vulnerable to ongoing negative effects from the stress condition. Only participants who scored less than 19 (thus fell into the category "likely to be well") were permitted to continue the experiment. Three potential candidates were rejected via this means.

All experimental stimuli were presented on a computer screen using custom software whilst participants were sitting comfortably at a desk. GSR signals were collected using GSR sensors from ProComp Infiniti GSR of Thought Technology Ltd. The sensors were attached to the non-dominant hand for all participants. GSR signals were sampled at a rate of 10Hz. Participants were asked to remain still and only move their dominant hand for mouse control during the experiment.

12.2.2 Procedures

The experiment consisted of a within-subjects, six-way factorial design. We administered math questions of three difficulty levels (low, medium and high) under two different stress conditions: "no-stress" and "stress". For level 1 problems (low difficulty), three terms were added together. For level 2 (medium difficulty), each problem consisted of four terms, with both addition and subtraction required. Level 3 problems (high difficulty) consisted of five terms, with addition, subtraction and multiplication required. The multiplication terms were in a random position within the problem [2].

All participants undertook the "no-stress" condition first. Participants were told that they would be completing math tasks but it was emphasized that their performance/accuracy was not important. After submitting some basic demographic information, a 2-min baseline period was carried out where the participants were told, via an on screen prompt, that they should just relax and let their mind wander. Then three 2-min blocks of math tasks were presented, each with 4 multiple choice answers. Tasks were not time limited and feedback was not provided. The three blocks in the "no-stress" condition were of level 1, 2 and 3 difficulty in sequential order. Participants were given a 2-min resting period in between each block [2].

After block three, the participants were asked, via on screen prompts, to nominate a "target score" for further tasks based on their estimation of their performance so far. Once submitted, the stress condition ensued. They were told that their performance would now be monitored. They were also informed of time limits for further trials, and video screens were switched on so that the test subject could see a video of themselves and also of other people observing them (see Fig. 12.1). Now in the stress condition, three more blocks of level 1, 2 and 3 math tasks were carried out, again with 2-min pauses in between each block but with time limits now imposed on each trial. Feedback ("correct", "wrong" or "out of time") was provided for 1 s after each trial.

Fig. 12.1 Experiment setup

12.2.3 Subjective Ratings

In order to validate the methods used in the experiment for inducing different levels of cognitive load, a one-way ANOVA test of pooled subjective ratings was conducted. Results showed that the disparity between the different difficulty levels was significant ($F = 82.32$, $p < 0.05$). It can be seen in Fig. 12.2 that the means of each group were increasing with the task difficulty level.

12.3 GSR Features Under Stress Conditions

12.3.1 Mean GSR Under Stress Conditions

The mean GSR values were inspected to study the effect of stress on cognitive load measurement. The distribution of normalised mean GSR values corresponding to the sub-sections of math task difficulty 1, 2 and 3 under both "no-stress" and "stress" conditions can be seen in Fig. 12.3. To investigate the relationship between mean GSR and cognitive load when no stressful stimuli are present, we conducted ANOVA analysis on these GSR values under "no-stress" conditions and found that there were statistically significant differences between the 3 different levels ($F = 10.5$, $p < 0.05$), and there is a noticeable upward trend in mean GSR that corresponds to an increase in task difficulty.

However, in the stress condition, the positive correlation between cognitive load and GSR could no longer be observed. Figure 12.4 shows the distribution of the

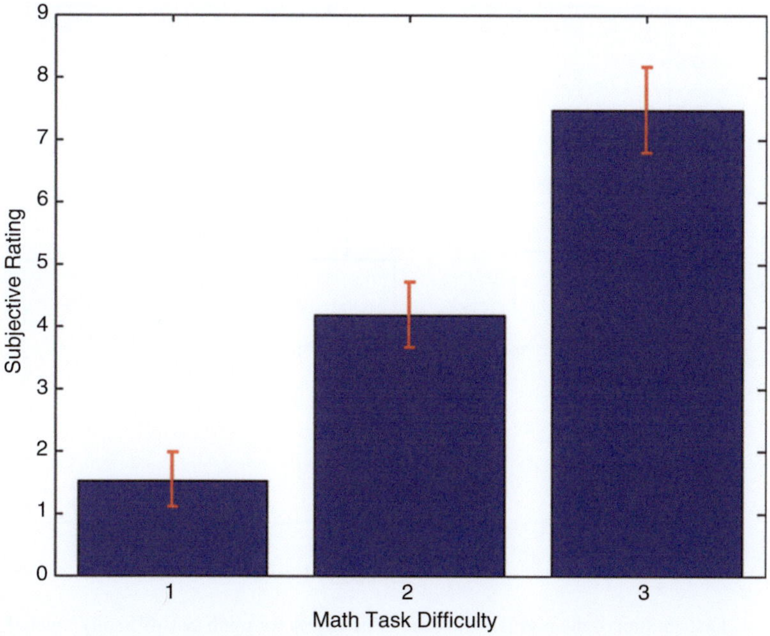

Fig. 12.2 Subjective rating of task difficulty

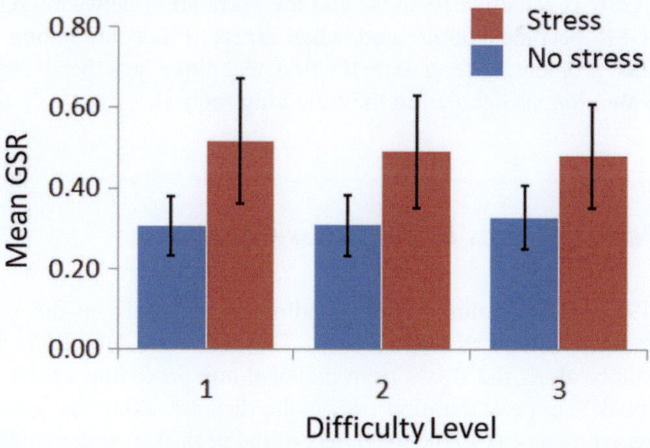

Fig. 12.3 Distribution of normalized mean GSR values for task difficulty levels 1, 2, 3 under the "no-stress" and "stress" conditions respectively

normalised mean GSR values for math task difficulty levels 1, 2 and 3, with both the "no-stress" and "stress" data included. ANOVA analysis of these values does not produce significant results ($F = 0.05$, $p = 0.95$). These results suggest that mean GSR cannot effectively index cognitive load when stress levels are fluctuating,

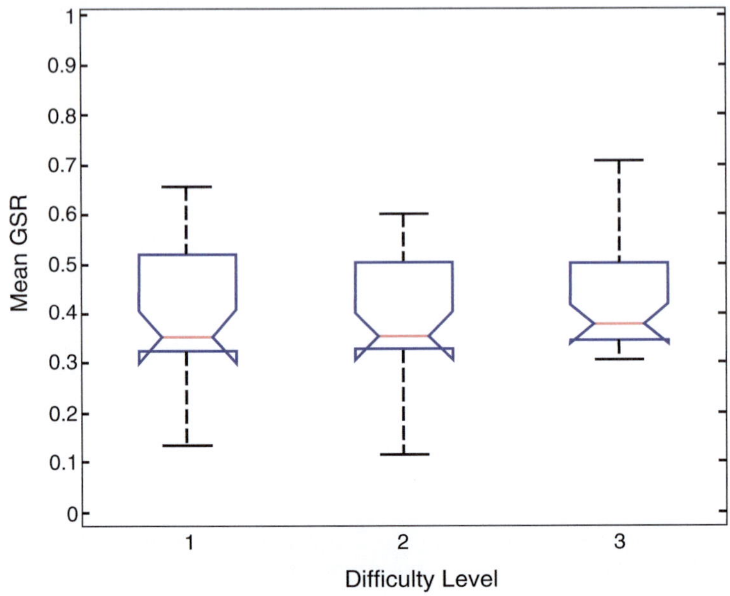

Fig. 12.4 Distribution of the normalised mean GSR values for math task difficulty levels 1, 2 and 3, with both the "no-stress" and "stress" data included

since mean GSR is sensitive to stress and the correlation between cognitive load and mean GSR becomes obfuscated when stress is a confounding factor. To overcome this problem, a feature extraction technique was been employed for workload evaluation, as detailed in the following section.

12.3.2 Peak Features Under Stress Conditions

Similar to [9], several features corresponding to the peaks in the signals were extracted from the smoothed GSR signals. The following definitions were made: S_D is the distance along the x-axis from the local min preceding a peak to the local max of the peak (i.e. peak duration); S_M is the distance along the y-axis from the local min preceding a peak to the local max of the peak (i.e. peak magnitude); S_F is the number of peaks divided by the task period (i.e. peak frequency). Figure 12.5 illustrates these concepts. In this work, these three peak based GSR features were used to study the effect of stress on the cognitive load measure. With the exception of S_M, the other two features, S_F and S_D, demonstrated their usefulness for indexing CL even when stress is a confounding factor.

The S_F feature represents the frequency of peaks per sub-section. The "no-stress" and "stress" distributions for this feature are shown in Fig. 12.6. For mean S_F, there was no common trend between the "no-stress" and "stress" conditions, and

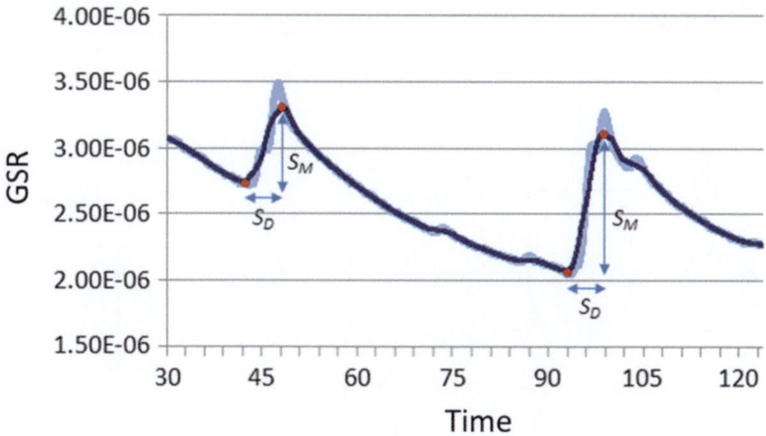

Fig. 12.5 Example of a smoothed GSR signal adorned with S_D and S_M features

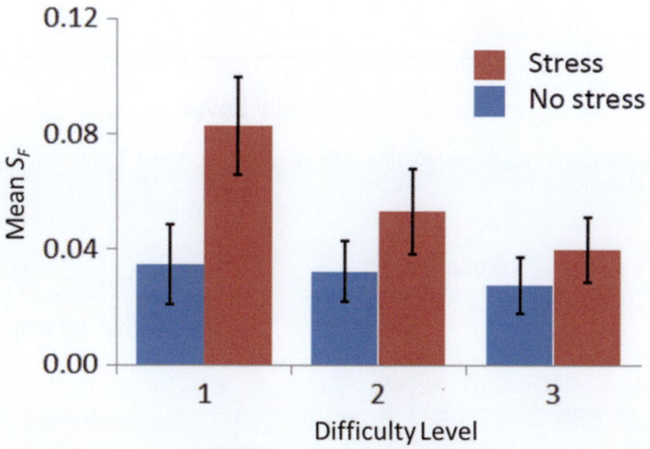

Fig. 12.6 Distribution of S_F feature for task difficulty levels 1, 2, and 3 under the "no-stress" and "stress" conditions

so there is no obvious way of using mean GSR to index CL when stress levels are fluctuating. However, we can see that S_F is negatively related to task difficulty regardless of whether stress is present, although the relationship is stronger in the "stress" condition. Figure 12.7 shows the distribution of the S_F feature for difficulty levels 1, 2 and 3, with both the "no-stress" and "stress" data included. ANOVA analysis was performed on this data to test the significance of the negative relation. The result exhibited significant difference among the three difficulty levels ($F = 3.96$, $p < 0.05$).

The S_D feature corresponds to the peak duration per sub-section. The distribution of the normalized S_D feature (sum of peak durations divided by the sub-section period) corresponding to math task difficulty 1, 2 and 3 under both "no-stress" and

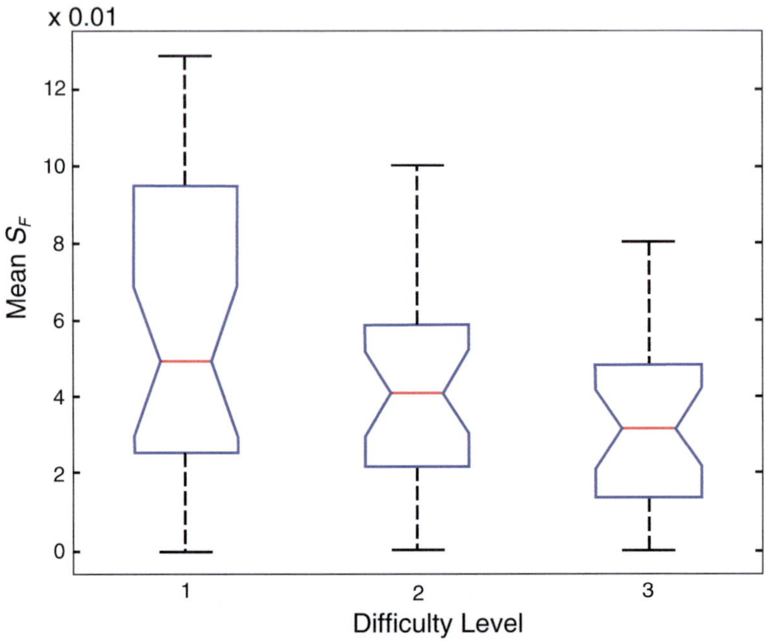

Fig. 12.7 Distribution of S_F feature for task difficulty levels 1, 2, and 3 with both "no-stress" and "stress" data included

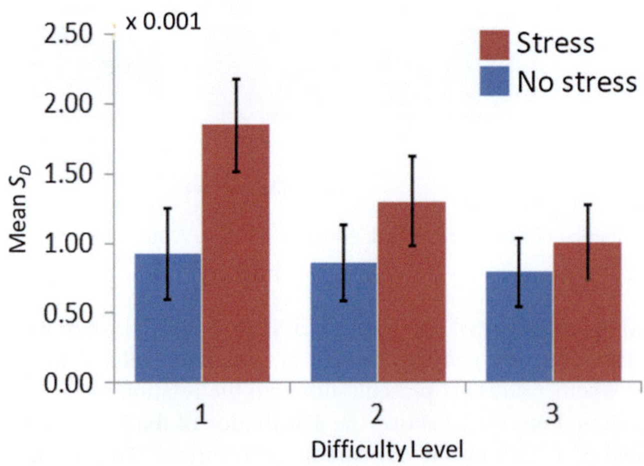

Fig. 12.8 Distribution of normalized S_D feature for task difficulty levels 1, 2, and 3 under the "no-stress" and "stress" conditions

"stress" conditions can be seen in Fig. 12.8. It turns out that this feature behaves quite similarly to the S_F feature, and is negatively related to task difficulty under both "no-stress" and "stress" conditions. Figure 12.9 shows the distribution of the S_D feature for difficulty levels 1, 2 and 3, with both the "no-stress" and "stress" data

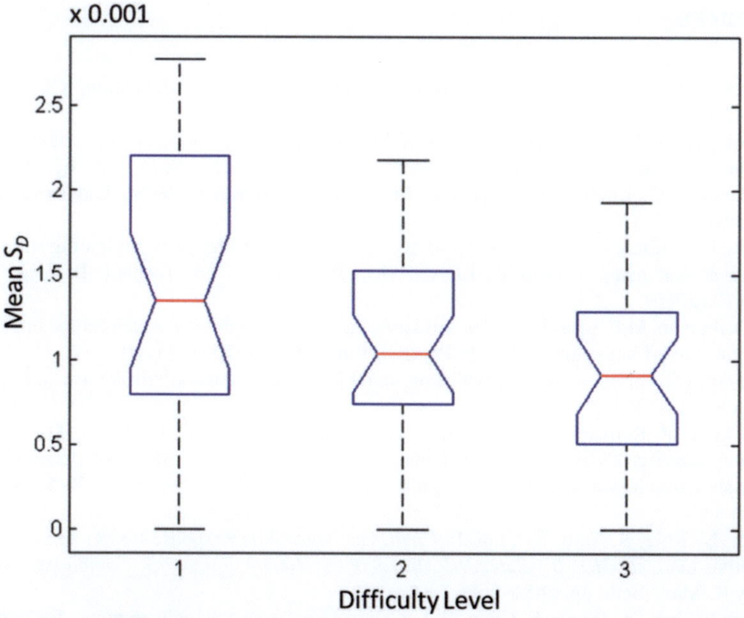

Fig. 12.9 Distribution of the S_D feature for difficulty levels 1, 2 and 3, with both the "no-stress" and "stress" data included

included. The downward trend with increasing task difficulty could be observed. ANOVA analysis also generated significant difference among the three difficulty levels ($F = 5.14$, $p < 0.05$), indicating the feature could be useful as an index of cognitive load even under the influence of stress conditions.

12.4 Summary

The study in this chapter helped to reinforce GSR as an index of cognitive load during task execution. Without the impact of stress, it appears that an increase in CL (induced by increasing the difficulty of tasks given to test subjects) results in an increase in mean GSR value. This relationship is, however, obfuscated when test subjects experience fluctuating levels of stress. Stress was introduced into the experiment using an adaptation of the MIST protocol, and this blurred the connection between GSR and CL.

GSR may still be useful as an index for CL even when stress is a confounding factor, if we consider peak based features extracted from the GSR signal other than the mean value. Both peak frequency in the signal and peak durations are negatively correlated to task difficulty and hence CL. These features could possibly be used to dissociate CL from stress and develop a stress-agnostic method of CL classification.

References

1. G.N. Martin, N.R. Carlson, W. Buskist, *Psychology*. Pearson Education, Old Tappan, USA (2007)
2. D. Conway, I. Dick, Z. Li, Y. Wang, F. Chen, The effect of stress on cognitive load measurement, in *Human-Computer Interaction – INTERACT 2013*, ed. by P. Kotzé, G. Marsden, G. Lindgaard, J. Wesson, M. Winckler (Springer, Berlin/Heidelberg, 2013), pp. 659–666
3. C. Setz, B. Arnrich, J. Schumm, R. La Marca, G. Tröster, U. Ehlert, Discriminating stress from cognitive load using a wearable EDA device. IEEE Trans. Inf. Technol. Biomed. **14**(2), 410–417 (2010)
4. S.S. Dickerson, M.E. Kemeny, Acute stressors and cortisol responses: A theoretical integration and synthesis of laboratory research. Psychol. Bull. **130**(3), 355–391 (2004)
5. H. Markus, The effect of mere presence on social facilitation: An unobtrusive test. J. Exp. Soc. Psychol. **14**(4), 389–397 (1978)
6. K. Dedovic, R. Renwick, N.K. Mahani, V. Engert, S.J. Lupien, J.C. Pruessner, The montreal imaging stress task: Using functional imaging to investigate the effects of perceiving and processing psychosocial stress in the human brain. J. Psychiatry Neurosci. **30**(5), 319–325 (2005)
7. Y. Shi, N. Ruiz, R. Taib, E. Choi, F. Chen, Galvanic Skin Response (GSR) as an index of cognitive load, in *CHI '07 Extended Abstracts on Human Factors in Computing Systems*, ed. by R. Mary Beth, pp. 2651–2656
8. N. Nourbakhsh, Y. Wang, F. Chen, R.A. Calvo, Using galvanic skin response for cognitive load measurement in arithmetic and reading tasks, in *Proceedings of the 24th Australian Computer-Human Interaction Conference*, New York, NY, USA, 2012, pp. 420–423
9. J. Healey, R. Picard, SmartCar: Detecting driver stress, in *Proceedings of 15th International Conference on Pattern Recognition 2000*, 2000, vol. 4, pp. 218–221
10. J. Zhai, A. Barreto, Stress detection in computer users through non-invasive monitoring of physiological signals. Biomed. Sci. Instrum. **42**, 495–500 (2006)
11. G. Andrews, T. Slade, Interpreting scores on the Kessler Psychological Distress Scale (K10). Aust. N. Z. J. Public Health **25**(6), 494–497 (2001)

Chapter 13
Trust and Cognitive Load

Trust has been found to be a critical factor driving human behavior in human-machine interactions with autonomous systems in modern complex high-risk domains such as aviation, and the military command and control [1, 2]. It is also one of the most important factors in management and organizational behavior for much of personal and business decision making as well as for efficiency and task performance [3]. Trust is associated with many variables, such as cooperation and monitoring behaviors, which can affect and reflect the extent of interpersonal trust. An understanding of how situations of higher cognitive load affect human's trust perception and vice versa is a critical factor for any organization in order to provide appropriate support to them and hence improve their overall task performance. A few studies have attempted to isolate the effects of high workload and stress on the level of human's trust judgments but their focus has traditionally been more on the human's trust perception of organizations and of automation systems. The highlights of this chapter include:

- Reviewing relations of trust and cognitive load;
- Presenting an experiment to find relations of trust and cognitive load;
- Investigating trust between human and information as well as interpersonal trust to find relations with cognitive load;
- Analyses of behavioral features such as linguistic features on trust and cognitive load.

13.1 Definition of Trust

Trust in autonomy is described as a multi-dimensional construct that changes with time. It is influenced by the types and format of information received by humans, their individual approaches to develop and determine trust, and aspects such as system capability and reliability [4]. One of the most widely cited definition of trust

F. Chen et al., *Robust Multimodal Cognitive Load Measurement*, Human–Computer Interaction Series, DOI 10.1007/978-3-319-31700-7_13

is from Lee and See [5], which defines trust as "the attitude that an agent will help achieve an individual's goals in a situation characterized by uncertainty and vulnerability". Trust is also defined as "one party's (or individual's) willingness to accept the vulnerabilities of actions or behavior of the other party based on the expectation that the other will perform the actions important to the trustor" [1, 6, 7]. Trustworthiness on the other hand is different from trust. Mayer et al. [1] found three trustworthiness elements that influence the development of trust in interpersonal situations: ability, benevolence, and integrity. Various approaches are proposed for trust measurement. For instance, it has been found that the extent of repeating chat abbreviations was high when communicators trusted their partners. Previous research also shows that there is an association between cognitive load and human's behavior, for instance, it was found that there was a high frequent use of complex sentences in speaking when cognitive load was increased [8]. It is significant to know how different trust factors during trust development and acquisition affect cognitive load and vice versa during task time.

13.2 Related Work

13.2.1 Trust

Mayer et al. clearly distinguished trust from other similar constructs like cooperation, confidence, and predictability [1]. Mayer argued that a person can still cooperate with someone even if he/she does not trust them because of some other external factors such as a fear of some kind of punishment. He also argued that prior risk recognition in case of trust is what differentiates it from confidence. Trust and predictability cannot be equated because a person cannot trust a party when the party is expected to constantly ignore others and act only for its own interests, just because the party is predictable [1].

Mayer also presented characteristics of a trustor and a trustee. The trustee's characteristics of ability, benevolence, and integrity, form the elements of trustworthiness. Benevolence is the perception of a positive orientation of the trustee toward the trustor [1, 7]. While on the other hand the trustor's propensity (or the expectation about the trustworthiness of others or general willingness to trust others) also affects one party's trust in the other party. Mayer proposed that trust is "a function of the trustee's perceived ability, benevolence, and integrity and of the trustor's propensity to trust". Mayer also emphasized that the level of trust (as determined by the three trustworthiness factors) will also be affected by the contextual factors.

Some studies argued that the relative effect of the elements of trustworthiness will vary depending on the type of activity to be performed and also by the person's cultural background [9]. For example, when a human is required to make some decisions or perform some actions based on his/her trust judgment about another human or entity, the most important factor in establishing the trust perception becomes the ability or competency of the other human or entity. While on the other hand, when a human is only required to make a judgment about another human or entity without the need to take some decision or perform some personal action based on that judgment, the most crucial factor in the trust perception is the integrity. For the third element of trustworthiness, benevolence, it is expected that it would be the most predictive factor in describing some feelings about a human. For example, when a human is asked about who he/she thinks would be a good friend, the most emphasized factor would the benevolence [9].

13.2.2 Trust and Cognitive Load

Cognitive load is a key component of the four-stage model of human information processing [10]. It is clear that, like trust, cognitive load plays an important role in mediating human behavior during collaboration with other humans or autonomous systems. Several behavioral changes may occur in human's natural work-related task performance due to varying levels of cognitive load induced by various task and non-task related factors. During high cognitive load, they might feel stressed or frustrated with the work situation, which may result in decreased task performance as well as lower level of trust perception for the human responsible for providing such a work environment. An understanding of how situations of higher cognitive load affect human's trust perception, therefore, is a critical factor for any organization in order to provide appropriate support to them and hence improve their overall task performance.

Despite their importance, little is known about the relationship between the two – trust and cognitive load – in such contexts. Intuitively we might guess that as cognitive load increases, a human may choose to rely more heavily on colleagues or an automatic system and in the process display more indicators of an implicit trust of the system. Alternatively, under higher load the human may be reticent to increase their trust and instead adapt their strategy to manage the increased task complexity to avoid increased dependence on others or on an automatic system. Despite the limited empirical evidence, Parasuraman and Riley [11] argued that increased workload is often cited as one of the most important factors in the choice to use automation. However, in a comparison of trust in various levels of automation, including manual control, Ruff and colleagues [12] found that as workload increased, subjective reports of trust decreased for automation, but increased for manual control. In an explanation for the equivocal evidence for trust and automation use, Parasuraman and Riley [11] suggested that complex task domains may prompt different task strategies, such as the use of automation during high cognitive load even if trust is low. Thus, under high cognitive load, the use of automation and trust (subjectively measured) may not be aligned, as is often assumed.

Previous investigation of the effect of cognitive load in user interfaces suggests the entrenchment of established behaviors with increasing load [12]. Other research also suggests that with increased cognitive load users revert to older, over learned and simpler types of responses [13]. When a user has a pre-existing trust of an automated system, the implication is that they will tend to over-trust the system during higher cognitive load [14], and a similar result might be expected of another human agent rather than a system. The Affect Infusion Model [15] suggests that during faster processing, humans use their affective states as a short-cut to infer their evaluative reactions to a target, such as in judgments of trust. Thus, it is likely that similar reliance on established cultural values and attitudes may rise to the surface during higher cognitive loads. With respect to the elements of trustworthiness, it has been found that action-oriented and performance-oriented cultures put more value on a party's ability, while collaborative and relationship-oriented cultures emphasize more on the human's benevolence.

A few studies have attempted to isolate the effects of high workload and stress on the level of human's trust judgments but their focus was more on the human's trust perception of organizations and of automation systems, eg [7] and [14]. The former study showed that when good performance is overlooked by an appraisal system in a job environment, employees develop a lack of trust in their employer. On the other hand, when they felt that the appraisal system was fair, their trust for top management increased and they regarded integrity as the most important factor for this trust perception. The study was based on the standard elements of trust and trustworthiness, ie, the ability, benevolence, integrity, and trustors' propensity as proposed by Mayer et al. in 1995. In the second example, Biros et al. [14] presented a study where the objective was to see how human's usage of and dependence on system automation (in other words their trust in a system) changes when they experience high task load, especially under information uncertainty situations. It was found that when task load (and hence cognitive load) increases, human continue to rely on the (interaction and decision support) system, even if they have less trust in it. In one study, the researchers found that under high load and critical task situations, human trust the system more when the system behaves in a polite manner and depicts accepted etiquette norms [16]. They also found that human show increased trust in the system they use when the system shows and maintains its reliability and dependability [16]. The study also discussed that as the system's automation becomes more complex (hence causing higher cognitive load), the humans' ability and willingness to learn and experience details of the system behavior also decreases and they tend to rely more on the system. The study also suggested that if we can make the system behave similar to a human the users find trustworthy, it will increase their level of trust in the system. The study concluded with the note that high task load along with inappropriate trust level can increase user's cognitive workload [16].

Both trust and cognitive load, in their individual capacity, are also known to affect the human behavior including their communication and/or linguistic behavior. For example, it has been found that less polite language or communication negatively affects human's trust, while a more polite language improves their trust perception about the communicator [1, 16]. Other studies found that under high trusting conditions, humans repeat each other's linguistic expressions or use similar words, ie they show high linguistic mimicry and similarity [17, 18]. Humans also use non-linguistic elements of speech like using more emphasis during their communication (ie high pitch and volume) when they trust more [19]. Like trust, cognitive load also affects human's linguistic behavior. Several recent studies have shown that a high cognitive load causes humans to show certain linguistic patterns in communication, especially when they work in a team environment, for example; increased use of pausing, hesitations, and self-corrections [20–22], increased use of negative emotions, more disagreement, decreased use of positive emotions, certain patterns of personal pronouns, and many other linguistic indicators of high cognitive load [23, 24] as presented in previous chapters of this book.

13.3 Trust of Information and Cognitive Load

In order to evaluate relations between trust and cognitive load, this section presents a user study to investigate the relationships between the concepts of trust, cognitive load, and human behavior, especially with respect to decision making. This section especially focuses on the human trust of information.

13.3.1 Task Design

A job applicant screening task was designed to be performed by a person under different cognitive load conditions with various trustworthiness elements embedded into the applicant's job profile. The objective was to assess how human's trust perception varied for human having various trustworthiness elements and under various complex cognitive load situations.

To produce higher cognitive load tasks, a dual-task paradigm was employed. Subjective ratings of complexity and difficulty were employed after each task set, to ensure that the desired levels of load built into the task design were actually being perceived by the study participants. The full set of all experimental conditions for a given factor can be seen in Table 13.1.

In terms of cognitive load measurement, the pair of high-trust tasks allowed a "control" condition, where variation was expected due only to high cognitive load that matches previously observed results (e.g. [25]. The task design ensured that multiple methods of cognitive load measurement were available, in particular, recordings of participants' speech and language and logged keystrokes/mouse movement (behavioral measure) as well as self-ratings (subjective measure). The study was designed with a 2 (cognitive load, w/n) $\times$ 3 (AIB indicators, w/n) mixed design.

All subjects completed both load conditions (low and high load) in a repeated measures design because the implicit load measures are dependent on a baseline to high load comparison. Expected changes in the recorded behavioral data features of

Table 13.1 Examination of interdependence between trust and cognitive load: experimental conditions

		Trust/trustworthiness	
		Low	High
Cognitive load	Low	Single task, low induced trust	Single task, high induced trust
	High	Dual task, low induced trust	Dual task, high induced trust

load will trend one way (e.g. increased pauses in speech during think-aloud of high load tasks).

The experimental platform simulated a computer-based applicant screening process called the "Human Resources Applicant Selection Tool". Participants were told that they were participating in a user evaluation of a new virtual interview tool being considered for a business application.

Participants assessed potential job candidates and reviewed the applicants' virtual resume which included standard experiential data (ie, education, previous experience, skills, etc.), interest statements, and referential data provided by previous supervisors. The aim was that the application would have a similar look and feel to that of Facebook or Linkedin. After several design discussions, story boarding and wireframe iterations, the final application was produced and is illustrated in Fig. 13.1. Candidate applicants' ability, integrity, and benevolence (AIB trustworthiness indicators) were manipulated through referential data inserted into the tool as well as through narratives provided. Each applicant was described by previous supervisors or co-workers as being high or low on one of the trustworthiness indicators. Examples of vignette-like descriptions of the trustworthiness indicators from previous research [26] were adapted for use in the current study. Four applicants were presented for each low or high load condition: 1 high benevolence candidate, 1 high ability candidate, 1 high integrity candidate and 1 neutral candidate on all three aspects. For each participant, a total of eight applicants were presented in both the conditions.

For the cognitive load manipulation, a secondary monitoring task was introduced, known as the notification feature, which allowed subjects to receive and "queue" new incoming resumes and applications to be "processed later". This was presented as an additional feature of the tool – and the participants were asked to complete two conditions of the task set, with and without the notification feature. The candidates provided in each condition were different instances, but represented the same AIB trustworthiness manipulations such that the entire task was exactly the same except for the notification feature as a dual task that resulted in the high cognitive load under that condition.

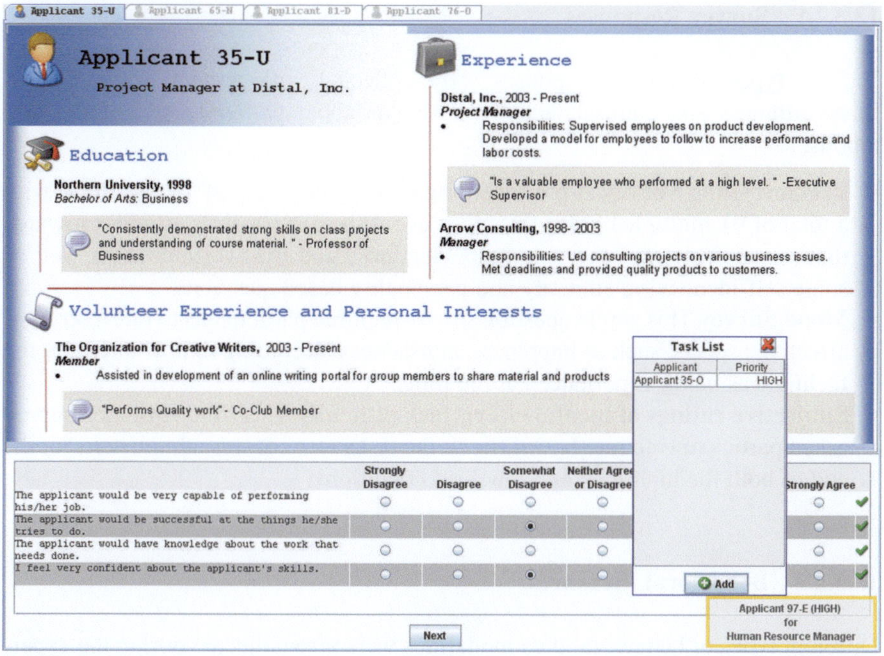

Fig. 13.1 The experimental platform used in the study

13.3.2 Data Collection

One hundred students from the INFO3315 course (Human-Computer Interaction) at a university participated in the experiment with both conditions (high and low CL). The experiment duration was approximately 1 h. Of these, 91 completed all sections of the study. The students received course credit in exchange for their participation, as well as snacks and movie ticket prizes for the "top performers" after the session.

A number of modalities and data streams were collected during this experiment. As mentioned before, the experiment was conducted employing a dual-task paradigm for higher cognitive load tasks. Subjective ratings of complexity and difficulty were employed after each task set, to ensure that the desired levels of load built into the task design were actually being perceived by the study participants. The experimental platform used in the study was developed in-house, and incorporated all data collection, in both versions (high CL and low CL). The following modalities of data were collected:

13.3.2.1 Survey Responses

Various types of survey responses were collected with respect to candidates' trustworthiness and cultural factors as well as about participants' attitudes and moods:

- **Pre-Screening Survey** A pre-screening survey consisting of 13 questions, with a total of 91 multiple choice questions about the participant's attitudes towards their supervisors and peers, honesty, kindness and trustworthiness, as well as some self-identifying ethnicity and personality based questions.
- **Mood Survey** This single question survey required participants to rate a series of affective aspects, such as happiness and sadness, according to how intensely the feeling was being experienced at the time.
- **Subjective ratings of mental effort/ task difficulty** This single question survey asked participants to rate how difficult the tasks were. It was administered at the end of both the high load and low load conditions.

13.3.2.2 Behavioral Measures

Different types of behavioral data modalities were also collected during the experiment including speech, text, and interactive data:

- **Speech data: think-aloud protocols** Participants were asked to verbalize their thought processes as they work through the three subtasks. Their speech utterances were recorded.
- **Speech Transcriptions** The speech data collected was transformed later to speech transcriptions for various task conditions as text data for linguistic analysis purpose.
- **Mouse trajectories** These were in the form of (x, y) coordinates, and were sampled with enough resolution to reproduce the entire experiment session. The mouse trajectory data was used to track widget manipulation and log use of the mouse movements or hovering over specific areas of the application. This data provided user's behavior in terms of their interaction with the application as well as an indication of their attentional focus.

13.3.2.3 Performance Measures

Different kinds of performance measures were also captured for the tasks performed during the experiment. These included:

- **Ratings, Filling positions and Rankings** The final responses to the actual subtasks.
- **Time-to-completion** Overall time to complete each task and the speech response time to answer each question.

13.4 Data Analyses

This chapter focuses on the linguistic feature analyses of speech transcription data collected during the study. Once the speech data was pre-processed and cleaned, mid-level features such as pause frequencies and lengths were annotated for pausing behavior analysis. The speech data was then manually transcribed and annotated for several linguistic features using a popular speech transcription and annotation tool called ELAN [27]. Once transcribed and annotated, several linguistic features were extracted from the transcriptions automatically using a text analysis tool called Linguistic Inquiry and Word Count (LIWC) [28]. LIWC comes with a built-in dictionary of various psycholinguistic category features comprising 4500 words. The LIWC dictionary comes with over 80 built-in word categories including negative emotion words, anxiety and anger words, cognitive processes, agreement and disagreement words, as well as grammatical parts of speech like personal pronouns and many others. The LIWC dictionary is customizable and hence two custom word categories relevant to trust analysis – Trust and Distrust words – were added to the dictionary and all trust and distrust words and their synonyms were included in the corresponding categories. The LIWC automatically extracted the linguistic features known to be relevant to trust and cognitive load [22–24] from the transcriptions. It extracted these features as percentages of total words spoken in order to deal with participants' verbosity differences. LIWC counts the number of words for a specific linguistic feature by matching the words from the transcription with its dictionary. The average dictionary coverage (the percentage of words captured by the dictionary) for the participants' transcriptions was over 95 %. The results of the linguistic analysis are presented in the results section. Table 13.2 lists the LIWC linguistic categories selected from the dictionary for analysis.

Table 13.2 Selected LIWC linguistic categories

Linguistic category	Example words
Negative emotions	ugly, nasty, bad, fail, sorry
Swear words	damn, shit, fuck
Anger	hate, kill, annoyed
Tentative	maybe, perhaps, guess
Certainty	always, never, absolutely
Achievement	won, done, performed
Trust	trust, believe, sure
Distrust	doubt, disbelieve, suspicious

13.5 Analysis Results

The analyses of participants' subjective ratings and speech transcription data were performed and the results are presented in this section.

13.5.1 Subjective Ratings of Mental Effort

To validate the experiment design for induced cognitive load levels, the subjective ratings of mental effort or task difficulty were collected from the participants. These were collected at the end of both the high load and low load tasks within each session and were based on a seven-point Likert scale (from 1 = "Extremely easy" to 7 = "Extremely difficult"). The analysis of the subjective ratings showed a mean rating of 3.625 for high cognitive load condition and 3.037 for low load condition as shown Fig. 13.2. The pair-wise statistical *t*-test showed a statistically significant difference between the ratings ($t(72) = 5.201, p < 0.001$). This confirmed an effective experiment design that induced the required levels of task difficulty and/or cognitive load levels as expected. The participants found the session with the dual-task more difficult overall than the session without the dual-task. This result is concurrent with other studies based on dual-task paradigm [29–32].

13.5.2 Linguistic Analysis of Think-Aloud Speech

The participants in the study were asked to verbalize their thought process using a "think-out-loud" protocol for all the tasks and their speech was recorded and transcribed using ELAN transcription and annotation tool. Out of 91 participants who completed both conditions, only 55 participants' speech was transcribed and

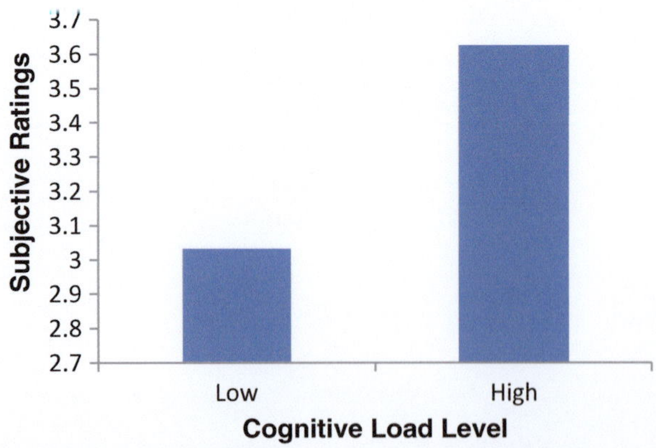

Fig. 13.2 Participants' subjective ratings for task difficulty ($p < 0.001$)

analyzed for linguistic behavioral changes for various trust and cognitive load features. The transcriptions were also annotated for various pausing and other mid-level behaviors like hesitations and repetitions. Hence, the linguistic analyses were carried out in three different areas – analysis of pausing behavior, analysis of LIWC-based linguistic categories, and analysis of other speech behavior.

13.5.2.1 Pause Analysis

Traditionally in psychology, the pauses during natural speech have been associated to a person's thinking and decision making, ie every time a person pauses during the speech, he/she processes currently known information in working memory to produce the next response [22]. Schilperoord also argued that the more time it takes to produce the response, the more cognitive energy is required to do so [33]. Accordingly for the current study, we hypothesized that participants will pause more and longer under high load situation than under the low load condition. We manually annotated different pause features mainly for two important factors of pauses – frequency and duration of pauses. The annotations were done in ELAN and covered various pause features as listed in Table 13.3.

Pauses inherently originate from breathing activity and are often very brief, so a cut-off value for pause lengths was defined [34]. Though selected arbitrarily, it usually ranges from 0.25 to 0.3 s [35]. For this study, it was assumed the threshold of 0.3 s; any pauses smaller than 0.3 s were assumed to be an inherent part of the natural speech and were not used for the analyses. As shown in Table 13.3, the overall pauses as well as silent and filled pauses were analyzed separately.

Paired-sample t-tests were performed and the results confirmed that overall, participants paused significantly more in the high load session than in the low load session ($p < 0.01$). Further investigation showed that both silent and filled pauses showed similar trends ($p < 0.01$). The overall pausing trends are illustrated in Fig. 13.3.

Table 13.3 Selected pause features and their brief description

Pause features	Description
Total pause frequency	Total number of pause segments
Freq. of silent pauses	Number of silent pause (voiceless) segments
Freq. of filled pauses	Number of filled pause (voiced) segments, e.g. ahh, um.
Avg. freq. of pauses/min	Average number of pauses per minute (normalized)
Avg. freq. of silent pauses/min	Average number of silent pauses per minute (norm.)
Avg. freq. of filled pauses/min	Average number of filled pauses per minute (norm.)
Total pause duration/length	Average length/duration of overall pauses (in seconds)
Length of silent pauses	Average length of silent pauses (in seconds)
Length of filled pauses	Average length of filled pauses (in seconds)
Percent of total time pausing	Percentage of total time spent in pausing (%)

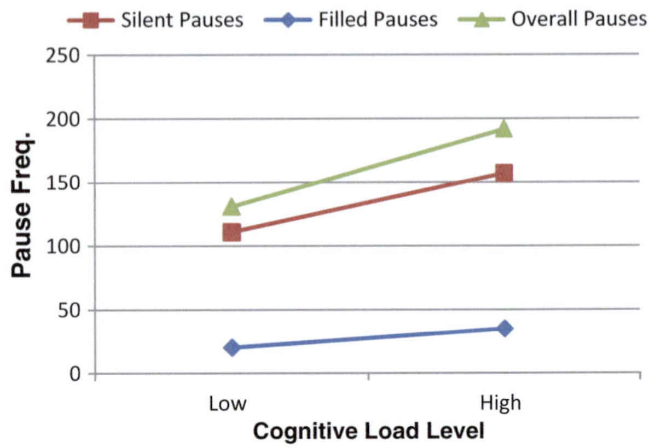

Fig. 13.3 Overall, silent, and filled pause frequencies ($p < 0.01$)

This pausing analysis was based on total number of pauses used by the participants and did not take the individual differences in speech into account. So to normalize any differences, average per minute pause frequencies were calculated for all participants and compared them for both low and high load sessions. It was found that participants still paused significantly more in high load condition than in low load situation ($p < 0.01$). The silent pauses also generated a similar significant result with more silent pauses under high load session ($p < 0.001$) as previously. The filled pauses, although still increasing under high load, did not show a significant difference with normalized frequencies ($p = 0.26$), which tells that the overall increasing pausing difference may have due to the silent pauses only and not the filled pauses. This also informed us that participants used significantly fewer filled pauses (<5) than silent pauses (~18) overall. These results are illustrated in Fig. 13.4.

The pause duration/lengths were also analyzed and paired-sample t-tests were performed on the overall pausing behavior under both sessions. The results confirmed that overall, participants paused significantly longer under the high load condition than the low load condition ($p < 0.0005$). Further investigation of silent and filled pauses showed that participants used longer silent pauses on the average under the high load condition than the low load condition ($p < 0.0008$) but there was no significant difference in the pause lengths for filled pauses ($p = 0.1$). The overall pausing duration behavior is illustrated in Fig. 13.5.

Finally, the overall percentage of total effective speech time the participants spent pausing was calculated for each participant, in order to normalize any other pausing differences across participants. The results confirmed that overall, participants spent more time pausing under the high load session as compared to the low load session ($p = 0.0001$), as expected. This is illustrated in Fig. 13.6.

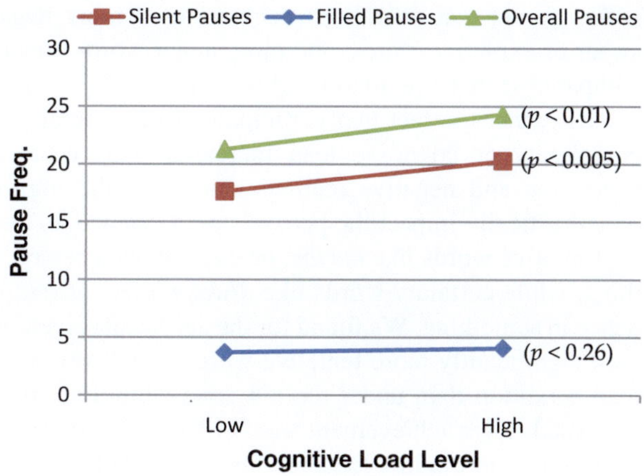

Fig. 13.4 Normalized pause frequencies

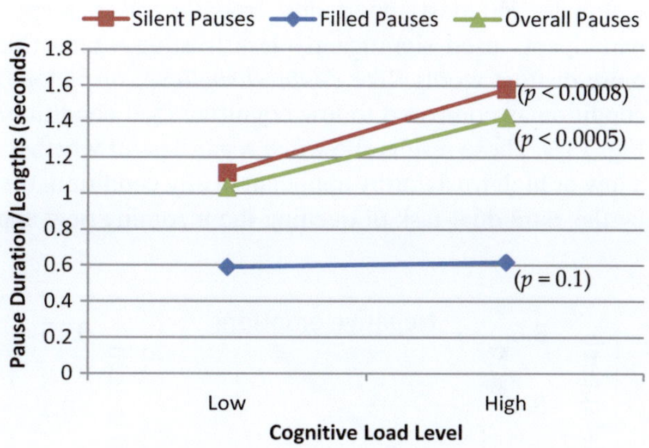

Fig. 13.5 Average pause duration/lengths (in seconds)

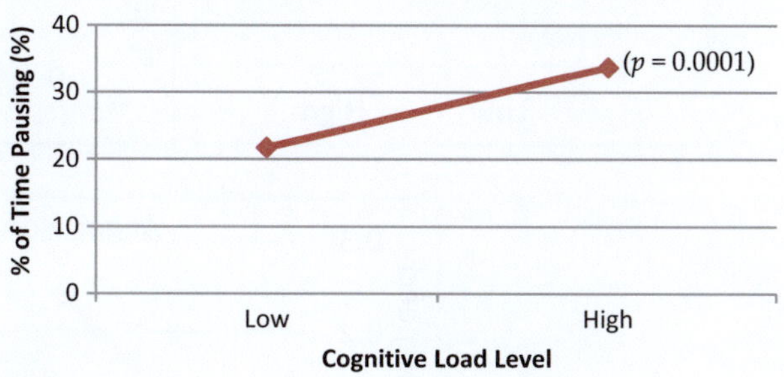

Fig. 13.6 Percentage of total speech time spent in pausing (%)

13.5.2.2 Linguistic Category Analysis

Apart from pausing features, some selected LIWC linguistic word categories as described in the previous subsection and presented in Table 13.2 earlier were also analyzed. Paired-sample t-tests were performed for these word categories and some interesting results were found with respect to their behavior under the two sessions.

It was found that overall participants used more negative emotion words, more swear or expletive words, and more anger words under the high load condition as compared to the low load condition ($p < 0.05$), as illustrated in Fig. 13.7. These results reflect the fact that participants feel more anger and frustration when task complexity or cognitive load increases. This is an important finding as high frustration and negative feelings caused by the high cognitive load of the task may drastically impact the person's performance on the task negatively.

Tentative words like *maybe, perhaps* show a person's doubtfulness about something, while certainty words like *always, sure, absolutely* depict a person's confidence in something. We found for the applicant screening study that the participants used significantly more tentative words and fewer certainty words under the high load condition than under the low load condition ($p < 0.05$). We also found that they used fewer achievement words like *won, done, performed* under the high load condition than the low load condition ($p < 0.05$), which also represents their poor task performance perception. These results are illustrated in Fig. 13.8.

Finally and most importantly, from the trust perception viewpoint, we found that participants used significantly less trusting words (like *trust, believe, sure*) and more distrust words (like *doubt, disbelieve, suspicious*) under high cognitive load condition as compared to low cognitive load condition ($p < 0.05$), as illustrated in Fig. 13.9. These results show that regardless of whether the participants were rating a low or high trustworthy applicant in any condition, the high cognitive load caused by the extra dual-task of queuing the incoming new applicants impacted their trust

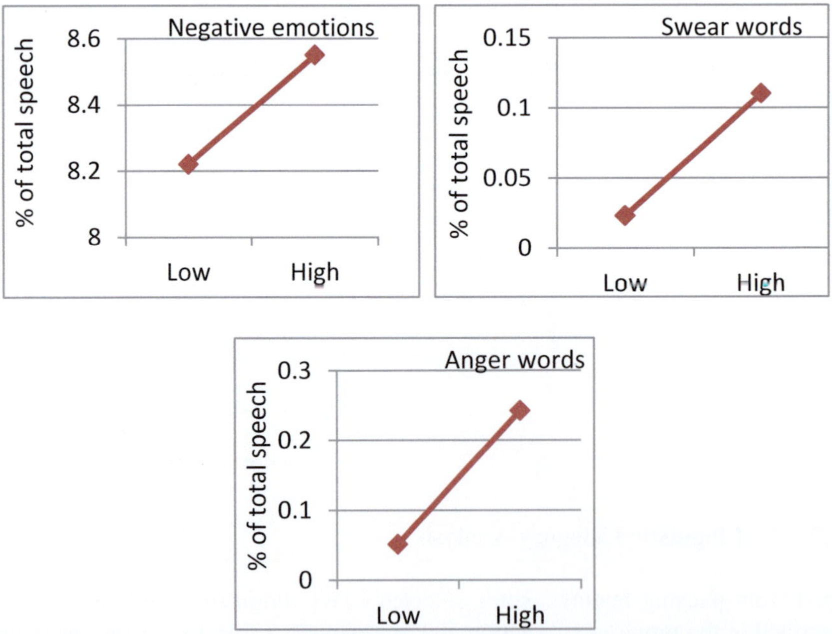

Fig. 13.7 Percentage of negative emotions, swear words, and anger words ($p < 0.05$)

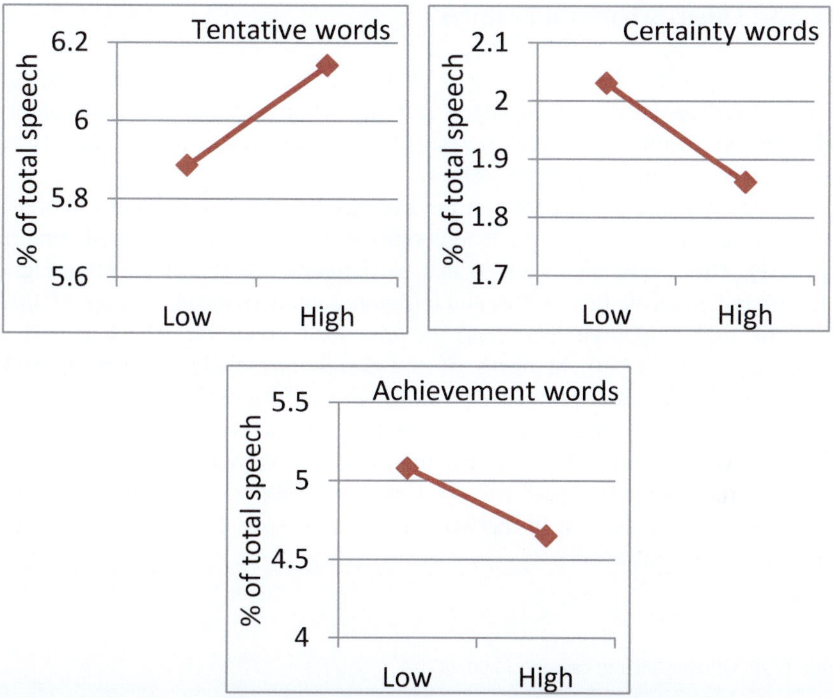

Fig. 13.8 Percentage of tentative, certainty, and achievement words ($p < 0.05$)

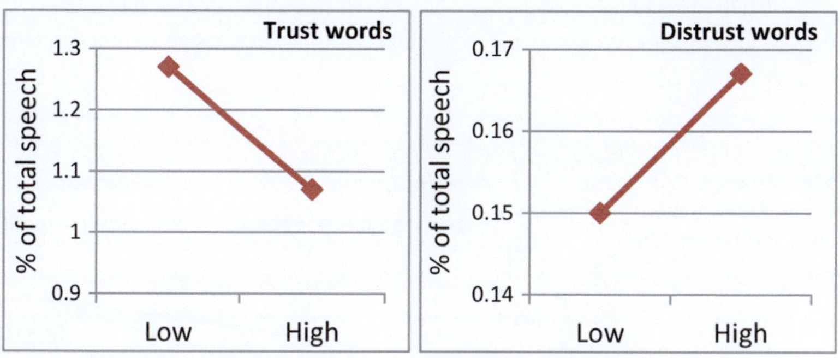

Fig. 13.9 Percentage of trusting and distrust words ($p < 0.05$)

perception negatively and changed their overall trust perception systematically. This finding is important as under highly critical task scenarios (e.g. in military operations, especially where personnel could be interacting with a sophisticated interaction system or a robotic system), a high cognitive load caused by task or interaction complexity could result in negative or lower trust perception by the person about the system being used, which in turn could possibly affect the person's task performance and/or impact their effective decision making ability.

13.5.2.3 Other Behavioral Features

Some other speech behavioral patterns including the number of times participants hesitated, repeated themselves, made self-corrections, and spoke in incomplete sentences (see Table 13.4) were investigated and analyzed for how they change under different cognitive load situations.

It was found that participants on the average hesitated significantly more frequently under the high cognitive load condition than under the low load condition ($p < 0.01$). The results also showed that participants repeated themselves significantly more frequently ($p < 0.001$) and left more sentences incomplete ($p < 0.0001$) under the high cognitive load than the low load condition. These results are illustrated in Fig. 13.10. In terms of self-corrections, there was an increasing trend but a significant difference ($p = 0.06$) was not found. These results suggest that high cognitive load affects participants' speech communication severely, which may result in some kind of miscommunication between the human working together collaboratively to perform some complex tasks and hence may reduce the overall task performance or in the worst cases may even threaten lives in critical scenarios like in war zones.

Table 13.4 Other speech behavioral features

Other behavioral features	Description
Avg. freq. of hesitations	Number of times they hesitate in their speech
Avg. freq. of repetitions	Number of times they repeat themselves
Avg. freq. of self-corrections	Number of times they correct themselves
Avg. freq. of incomplete sentences	Number of times they speak incomplete sentences

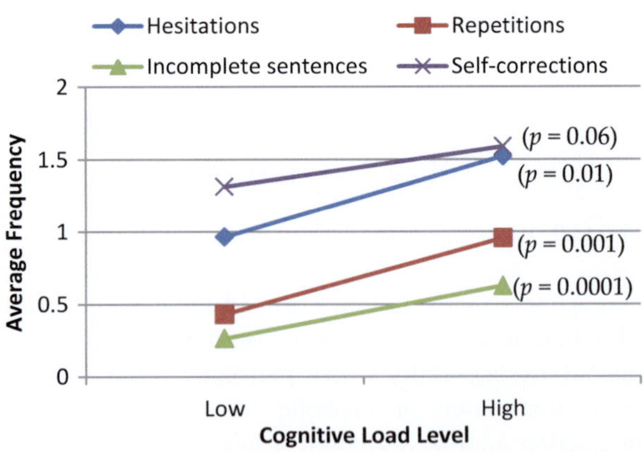

Fig. 13.10 Average frequency of hesitations, repetitions, and incomplete sentences

13.6 Interpersonal Trust and Cognitive Load

The previous sections investigated the relations of trust and cognitive load, especially the trust between human and information. This section presents how interpersonal trust can be affected by two types of cognitive load, low and high. It was expected that the establishment of trust becomes worse with a high cognitive load. These data sets were same data used and published in [36] to examine trust and cognitive load.

13.6.1 Task Design

The method that was used to investigate the level of trust is explained in detail in Sect. 9.2.2 in Chap. 9. Using this method, twenty participants were partnered randomly and chatted via instant messaging. During their communication, they faced a low cognitive load for 15 min (by summing small numbers) and a high cognitive load for 15 min (by summing large numbers). Pop-up boxes containing the numbers to be summed appeared on each participant's chat screen equally across the time. Also, as detailed in Sect. 9.2.2 in Chap. 9, the responses to the questionnaire indicated that the participants felt they faced a low load when the numbers to be summed were small and a high cognitive load when the numbers to be summed were large [36].

Interpersonal trust was measured using a questionnaire which included several questions (with responses to be given on a Likert scale from 1 to 9 for each question) about the extent of the participants' trust. These questions were taken from [37] and were modified for the purposes of this study. Specifically, the questions covered the attitudes of the participants toward their partners where, in addition to the text-chat communication, they were also required to play an investment game with their partners, the goal being to invest money and make a profit (more details on the investment game are given in Sect. 9.2.2 in Chap. 9). Examples of the questions in the questionnaire are: "*I feel my partner did not take advantage of me to maximize his/her outcome money.*" and "*I usually know how much he/she will invest in each round.*"

13.6.2 Results

The participants' trust in their partner was higher when they faced a low cognitive load, as indicated by the responses to the questionnaire on the Likert scale where the average of the Likert scale responses was 6.28 whereas the average of the Likert scale responses of the participants under a high cognitive load was only 4.98 (see Sect. 9.2.2) [36]. A two-tailed *t*-test was used to compare the responses of the high

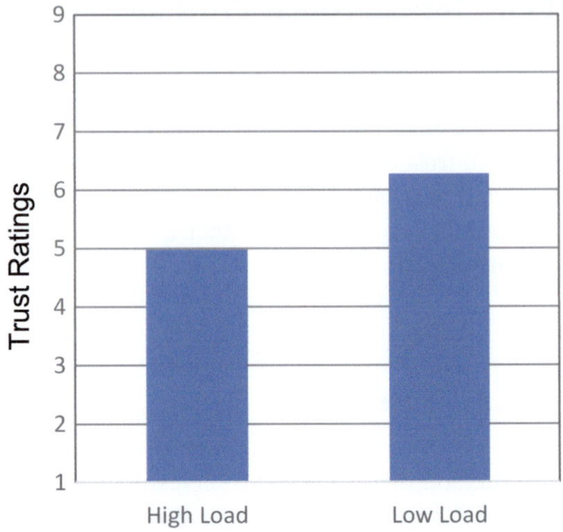

Fig. 13.11 The average of the Likert scale responses to the questionnaire on measuring trust (9 indicates high trust) [36]

and low cognitive load groups, which showed there was a significant difference between them ($p < 0.05$) (see Fig. 13.11) [36].

The degree of trust which was established between the participants during the 30-min communication sessions was affected by the level of cognitive load, that is, trust increased with a low cognitive load [36]. It is well known that cognitive load has an effect on people' behaviors, and the findings reported in this chapter establish that there is an association between interpersonal trust and cognitive load and furthermore shows that there are positive and negative correlations between them.

These findings open a window of opportunity for specialist researchers, such as those in the field of psychology, to further investigate the effects of the correlation between cognitive load and trust on people in various fields. Also, these findings may be important when users interact with systems and software [36]. For instance, if systems and software are easier to use, then users may have more trust and consequently this may also encourage developers to consider cognitive load as an important factor in their designs.

13.7 Summary

This chapter investigated the relations between trust perception and cognitive load. An experimental platform was designed and employed to collect multimodal data. A detailed data analysis and results were presented and discussed with the aim of understanding the relationships between cognitive load and trust. Analyses of speech transcription data and subjective ratings were conducted to understand relationships between cognitive load and trust. The results showed participants' varying behavioral indicators under different levels of cognitive load.

References

1. R. Mayer, J.H. Davis, F.D. Schoorman, An integrative model of organizational trust. Acad. Manag. Rev. **20**, 709–734 (1995)
2. D. Schmorrow, K. Stanney, *Augmented Cognition: A Practitioner's Guide* (Human Factors and Ergonomics Society Press, Santa Monica, 2008)
3. T. Donaldson, The ethical wealth of nations. J. Bus. Ethics **31**, 25–36 (2001)
4. A. Finn, Report on trusted autonomy workshop February 2014, University of South Australia, Mawson Lakes, SA, Australia, Technical Report DASI-2014-CR-005, Mar. 2014
5. J.D. Lee, K.A. See, Trust in automation: Designing for appropriate reliance. Hum. Factors **46** (1), 50–80 (2004)
6. J.R. Dunn, M.E. Schweitzer, Feeling and believing: The influence of emotion on trust. J. Pers. Soc. Psychol. **88**(5), 736–748 (2005)
7. R.C. Mayer, J.H. Davis, The effect of performance appraisal system on trust for management: A field quasi-experiment. J. Appl. Psychol. **84**(1), 123–136 (1999)
8. M.A. Khawaja, F. Chen, N. Marcus, Measuring cognitive load using linguistic features: Implications for usability evaluation and adaptive interaction design. Int. J. Hum. Comput. Interact. **30**(5), 343–368 (2014)
9. J.B. Lyons, C.K. Stokes, Exploring a dynamic model of trust management, *A white paper by Air Force Research Laboratory*, 2010
10. F. Paas, J.E. Tuovinen, H. Tabbers, P. Gerven, Cognitive load measurement as a means to advance cognitive load theory. Educ. Psychol. **38**(1), 63–71 (2003)
11. R. Parasuraman, V. Riley, Humans and automation: Use, misuse, disuse, abuse. Hum. Factors: J. Hum. Factors. Ergonomics. Soc. **39**(2), 230–253 (1997)
12. H. Ruff, S. Narayanan, M. Draper, Human interaction with levels of automation and decision-aid fidelity in the supervisory control of multiple simulated unmanned air vehicles. Presence: Teleoperators. Virtual. Environ. **11**(4), 335–351 (2002)
13. S. Oviatt, R. Coulston, R. Lunsford, When do we interact multimodally? Cognitive load and multimodal communication patterns, in *ICMI '04: Proceedings of the 6th international conference on Multimodal interfaces*, New York, NY, USA, 2004, pp. 129–136
14. D.P. Biros, M. Daly, G. Gunsch, The influence of task load and automation trust on deception detection. Group Decis. Negot. **13**, 173–189 (2004)
15. J. Forgas, Mood and judgment: The affect infusion model (AIM). Psychol. Bull. **117**, 39–66 (1995)
16. R. Parasuraman, C.A. Miller, Trust and etiquette in high-criticality automated systems. Commun. ACM **47**(4), 51–55 (2004)
17. L.E. Scissors, A.J. Gill, K. Geraghty, D. Gergle, In CMC we trust: The role of similarity, in *Proceedings of the 27th international conference on Human factors in computing systems*, 2009, pp. 527–536
18. L.E. Scissors, A.J. Gill, D. Gergle, linguistic mimicry and trust in text-based CMC, in *Proceedings of the 2008 ACM conference on Computer supported cooperative work*, 2008, pp. 277–280
19. B. Waber, M. Williams, J. Carroll, A. Pentland, A voice is worth a thousand words: The implications of the micro-coding of social signals in speech for trust research, in *Handbook of Research Methods on Trust*, ed. by F. Lyon, G. Möllering, M. Saunders (Edward Elgar, Cheltenham/Northampton, 2012), pp. 249–258
20. A. Berthold, A. Jameson, Interpreting symptoms of cognitive load in speech input., in *Seventh International Conference on User Modeling (UM99)*, 1999
21. A. Jameson, J. Kiefer, C. Müller, B. Großmann-Hutter, F. Wittig, R. Rummer, Assessment of a user's time pressure and cognitive load on the basis of features of speech, in *Resource-Adaptive Cognitive Processes* (Springer, Berlin/Heidelberg, 2009), p. 171

22. M.A. Khawaja, N. Ruiz, F. Chen, Think before you talk: An empirical study of relationship between speech pauses and cognitive load, in *Australasian Computer-Human Interaction Conference (OzCHI'08)*, Cairns, Australia, 2008, pp. 335–338
23. M.A. Khawaja, F. Chen, N. Marcus, Using language complexity to measure cognitive load for adaptive interaction design, in *Proceedings of International Conference on Intelligent User Interfaces (IUI 2010)*, Hong Kong, China, 2010, pp. 333–336
24. M.A. Khawaja, F. Chen, N. Marcus, Analysis of collaborative communication for linguistic cues of cognitive load. Int. J. Hum. Factors. Ergonomic. Soc. **54**(4), 518–529 (2012)
25. Y. Shi, N. Ruiz, R. Taib, E. Choi, F. Chen, Galvanic Skin Response (GSR) as an index of cognitive load, in *Proc. SIGCHI Conference on Human Factors in Computing Systems (CHI'07)*, San Jose, 2007, pp. 2651–2656
26. R.C. Mayer, P.M. Norman, Exploring attributes of trustworthiness: A classroom exercise. J. Manag. Educ. **28**(2), 224–249 (2004)
27. ELAN: (EUDICO Linguistic Annotator). Max Planck Institute for Psycholinguistics, The Netherlands, Apr-2010
28. J.W. Pennebaker, C.K. Chung, M. Ireland, A. Gonzales, R.J. Booth, The development and psychometric properties of LIWC2007, *www.liwc.net*, last accessed: April 2012, 2007
29. E. Leyman, G. Mirka, D. Kaber, C. Sommerich, Cervicobrachial muscle response to cognitive load in a dual task scenario. Ergonomics **47**(6), 625–645 (2004)
30. N. Marcus, M. Cooper, J. Sweller, Understanding instructions. J. Educ. Psychol. **88**, 49–63 (1996)
31. J. Sweller, J. Merrienboer, F. Paas, Cognitive architecture and instructional design. Educ. Psychol. Rev. **10**(3), 251–296 (1998)
32. F. Wada, M. Iwata, S. Tano, Information presentation based on edtimation of human multimodal cognitive load, *Proceedings of IFSA World Congress and 20th NAFIPS International Conference*, vol. 5, pp. 2924–2929, 2001
33. J. Schilperoord, On the cognitive status of pauses in discourse production, in *Contemporary Tools and Techniques for Studying Writing* (Kluwer Academic Publishers, London, 2001)
34. H.W. Dechert, M. Raupach, *Towards a Cross-Linguistic Assessment of Speech Production* (Lang, Frankfurt, 1980)
35. J. Schilperoord, *It's About Time: Temporal Aspects of Cognitive Processes in Text Production* (Rodopi, Amsterdam/Atlanta, 1996)
36. A. Khawaji, F. Chen, J. Zhou, N. Marcus, Trust and cognitive load in the text-chat environment: The role of mouse movement, in *Proceedings of the 26th Australian Computer-Human Interaction Conference on Designing Futures: The Future of Design*, New York, NY, USA, 2014, pp. 324–327
37. J.K. Butler, Toward understanding and measuring conditions of trust: Evolution of a conditions of trust inventory. J. Manag. **17**(3), 643–663 (1991)
38. D. Schmorrow, K. Stanney, *Augmented Cognition: A Practitioner's Guide* (Human Factors and Ergonomics Society Press, Santa Monica, 2008)

Part V
Making Cognitive Load Measurement Accessible

Chapter 14
Dynamic Cognitive Load Adjustments in a Feedback Loop

Cognitive load has been found to be a critical factor driving human behavior in human-machine interactions in modern complex high-risk domains. By monitoring a user's state and adapting the task difficulty levels, a dynamic cognitive load system can improve a user's performance and helps users maximize their capacity for productive work [1]. The highlights of this chapter include:

- Presenting a dynamic cognitive load adjustment feedback loop with a dynamic cognitive load adaptation model to control workload adjustment during human-machine interaction;
- An arithmetic addition task is used to show how task difficulty levels are dynamically adapted based on cognitive load measurement;
- Physiological signals such as GSR are utilized to obtain passive human sensing data in the presented model;
- By analyzing the obtained sensing data in real-time, the task difficulty levels are adaptively adjusted to better fit the user during working time;
- The presented feedback loop helps to balance the task performance and work-load levels.

14.1 Dynamic Cognitive Load Adjustments

The cognitive load experienced by a user in completing a task has a major impact on her/his ability to acquire information during the task, and can severely influence their overall productivity and performance. High levels of cognitive load are known to decrease effectiveness and performance of users, as well as their ability to learn from their tasks [2]. On the other hand, if a task is very easy and routine, inducing only a low level of cognitive load, it may cause boredom and loss of focus, ultimately resulting in lower performance. In this way, the concept of an optimal range of cognitive load levels is developed, outside of which a subject's ability to

F. Chen et al., *Robust Multimodal Cognitive Load Measurement*, Human–Computer Interaction Series, DOI 10.1007/978-3-319-31700-7_14

learn, perform, and complete a task is likely to be negatively affected [3]. It is crucial to maintain the cognitive load experienced by a user within this optimal range to achieve the highest productivity.

In order to keep the user in an optimal state and improve their engagement and performance, Dynamic Workload Adjustment (DWA) systems automatically modulate the difficulty of tasks and other factors related to tasks in human-machine systems in real-time. By monitoring a user's state and adapting the task difficulty levels, a dynamic workload system improves a user's performance and helps users maximize their capacity for productive work. Afergan et al. [4] used functional Near-InfraRed Spectroscopy (fNIRS) to detect task difficulty and optimize workload with a dynamic adaptation. Koenig et al. [5] presented a closed-loop control of cognitive load in neurological patients during robot-assisted gait training. Moreover, as investigated in previous chapters, physiological signals such as heart rate, breathing frequency, skin conductance, pupillary dilation and skin temperature were recorded for CL estimations. GSR has attracted researchers' attention as a prospective physiological indicator of cognitive load [6]. For example, Son and Park [7] estimated a driver's cognitive load using driving performance and skin conductance level as well as other measures in a driving simulator. The results showed that the skin conductance level provides clear changes associated with the difficulty level of the cognitive workload. Nourbakhsh et al. [6] also indexed cognitive load with GSR features of accumulative GSR and GSR power spectrum in arithmetic tasks. Wang et al. [8] indexed cognitive load with GSR features such as mean-difference. GSR data were also tested by the Boosting algorithm with Haar-like features for cognitive load classifications. This chapter investigates the use of GSR in a dynamic workload adaptation feedback loop in order to improve task performance.

14.2 Dynamic Workload Adaptation Feedback Loop

A dynamic workload adaptation feedback loop is proposed as shown in Fig. 14.1 [1]. In this feedback loop, physiological signals such as GSR are recorded when the user is performing a task. The recorded signals are then analyzed and classified as different CL levels. The classified CL levels are input into the adaptation model in order to modulate task elements. A new task session is then performed based on the adaptation in order to keep task difficulty on an optimal level for the participants.

14.2.1 Task Design

An arithmetic addition task was used in this study. Each task was designed to stimulate a particular CL level for the participant. An "X" was shown at the beginning on a computer LCD display followed by four numbers in succession,

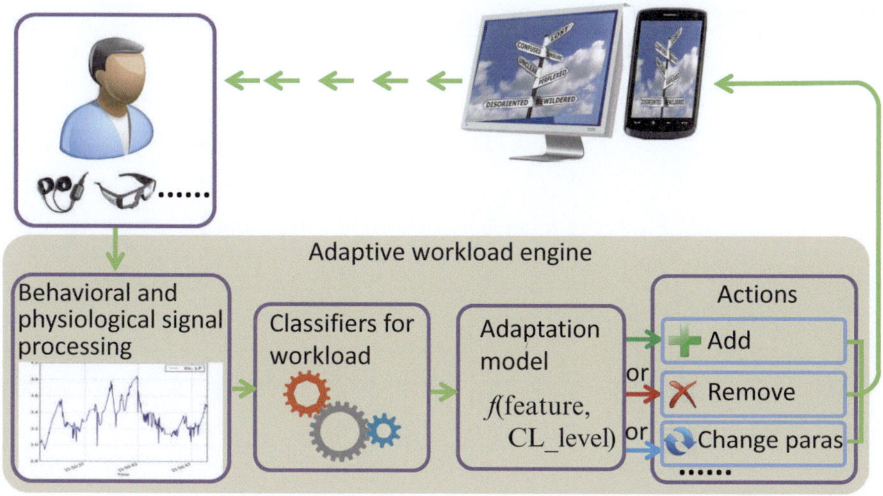

Fig. 14.1 Diagram of feedback loop of dynamic workload adaptation

where each number and "X" were displayed for 3 s. The participant was required to add these numbers up in his/her head during the task, and had to choose one answer from four options. At the completion of each task, the collected data were analyzed by the system. A new task is followed with a particular CL level controlled by the adaptation model. CL levels are designed as in Table 14.1 based on [6], where the number in each of the columns represents how many of the particular category of number (binary, 1-digit, 2-digit, 3-digit) are shown in the task.

14.2.2 Procedures

The experiment was carried out with the following procedures: (1) A computer was firstly setup with the GSR sensor connected and the corresponding drivers installed. (2) The participant was seated facing the LCD display of the computer. The tips of the index and fourth fingers of the left hand of the participant were connected to the GSR sensor, and the right hand of the participant was used to navigate the mouse to engage with the program. (3) The experiment began by launching the experiment program. (4) The training stage firstly ran for a total of 8 tasks. (5) After training was completed, 6 testing stages are run, and each testing stage had a total of 20 tasks. Table 14.2 illustrates the different testing stage scenarios. The numbers displayed for the arithmetic tasks were all randomized, and the difficulties of tasks were controlled by the adaptation model. GSR devices from ProComp Infiniti GSR of Thought Technology Ltd were used in the experiment. 10 participants of university students and research staff were recruited in this experiment.

The scenarios of testing stages were as follows:

Table 14.1 Cognitive load level definitions

CL	Binary	1-Digit	2-Digit	3-Digit
0	4	0	0	0
1	3	1	0	0
2	2	2	0	0
3	1	3	0	0
4	0	4	0	0
5	0	3	1	0
6	0	2	2	0
7	0	1	3	0
8	0	0	4	0
9	0	0	3	1
10	0	0	2	2
11	0	0	1	3
12	0	0	0	4

Table 14.2 Testing stage scenarios

Testing stage	Initial CL	Adaptation model
1	0	1
2	6	1
3	12	1
4	0	2
5	6	2
6	12	2

- The testing stages starting from an initial CL level of 0 and 12 were designed to test the effectiveness of the adaptation model in allowing the difficulty level to shift from one extreme to a more desirable state.
- The testing stages starting from an initial CL level of 6 were designed to test the robustness of the adaptation model, and observe its effectiveness in keeping the CL level stable at the desirable state.
- Two different adaptation models were used for comparison resulting in 6 experiment scenarios.

14.3 GSR Features

14.3.1 Signal Processing

The raw GSR signals were firstly calibrated in order to account for variations of GSR between individuals and time intervals. As mentioned in Sect. 14.2.1, an “X” was shown at the beginning before the actual arithmetic task begins. This period was used as the reference point on which the rest of the data could be calibrated. The calibration was achieved using the relationship:

$$G_T = \frac{G_t - G_X}{G_X}$$

where G_X is the average GSR value during the X-displaying period, and G_t and G_T are the raw and calibrated GSR signals during the task time respectively. Signal smoothing was achieved using a Hann window function as a low pass filter to remove high frequency noise.

14.3.2 Feature Extraction

Time domain features and frequency domain features were extracted in this study. For the time domain features, we focused on analyzing the nature of the major peaks in the data. In order to normalize the magnitudes of these peaks across all participants, the processed GSR signal for each task was divided by the mean value of all tasks of the particular participant:

$$G_N(i,k,t) = \frac{G_T(i,k,t)}{\frac{1}{m}\sum_{j=1}^{m}\sum_{t=1}^{T_{ij}} G_T(i,j,t)}$$

where $G_T(i, k, t)$ is the result from signal processing and $G_N(i,k,t)$ is the normalised GSR value at time t of task k of subject i.

A peak was identified as the significant peak by setting thresholds for both the time period S_{d_i} and height S_{m_i} of the peak (see Fig. 14.2). As shown in Fig. 14.2, point 1 is not considered as a peak as its time duration and height do not reach the threshold value, and point 2 is considered as a significant peak.

The significance of each peak i is quantified using the duration of the peak S_{d_i}, magnitude S_{m_i}, and area $S_{a_i} = S_{d_i}S_{m_i}$. The time domain features include: (1) Sum of peak durations $S_d = \sum S_{d_i}$; (2) Sum of peak magnitudes $S_m = \sum S_{m_i}$; (3) Sum of peak areas $S_a = \sum S_{a_i}$; (4) Number of peaks S_f; and (5) Time taken to choose answer T_c.

For frequency-domain feature extraction, Z-score normalisation was firstly applied:

$$G_{Z(i,j)} = \frac{G_T(i,j) - \mu_{G_T(i,j)}}{\sigma_{G_T(i,j)}}$$

where G_T is the calibrated and smoothened GSR signal and G_Z is the normalised signal of task j of subject i. μ and σ are the mean and standard deviation of $G_T(i,j)$. Since each task was normalised in this way using its own mean and standard deviation, the magnitudes and range of G_Z of each task became standardised, thus the frequency features could be isolated more effectively. The power spectrum was extracted using:

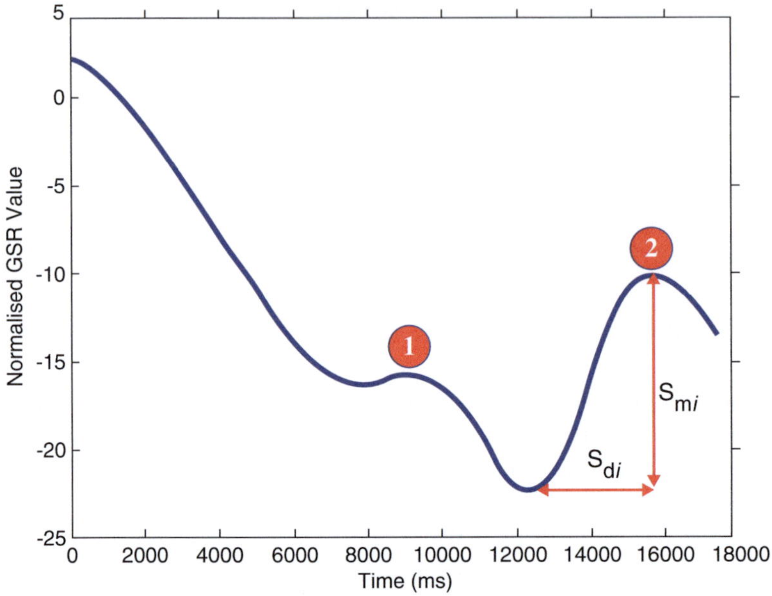

Fig. 14.2 Time-domain feature extraction

$$P(\omega) = \frac{1}{N} Y(\omega) Y^*(\omega)$$

where P is the power spectrum, ω is frequency, N is the length of the signal, and Y and Y^* are the frequency spectrum and its complex conjugate respectively. The average power below 1Hz was calculated for each task, as this frequency region contained the most non-zero values.

In order to optimise the performance of the machine learning algorithms, the extracted features were then normalized to a Z-score, so that each feature would have a zero mean and a standard deviation of 1. For any particular feature $f(j)$ in task j, the normalised $f_z(j)$ is calculated using the following equation:

$$f_z(j) = \frac{f(j) - \mu_f}{\sigma_f}$$

where μ_f is the mean of features $f(j)$, σ_f is the standard deviation of features of $f(j)$.

14.4 Cognitive Load Classification

14.4.1 *Offline Cognitive Load Classifications*

Using all of the features mentioned above, three different Machine Learning (ML) methods were used to classify CL levels: Support Vector Machine (SVM),

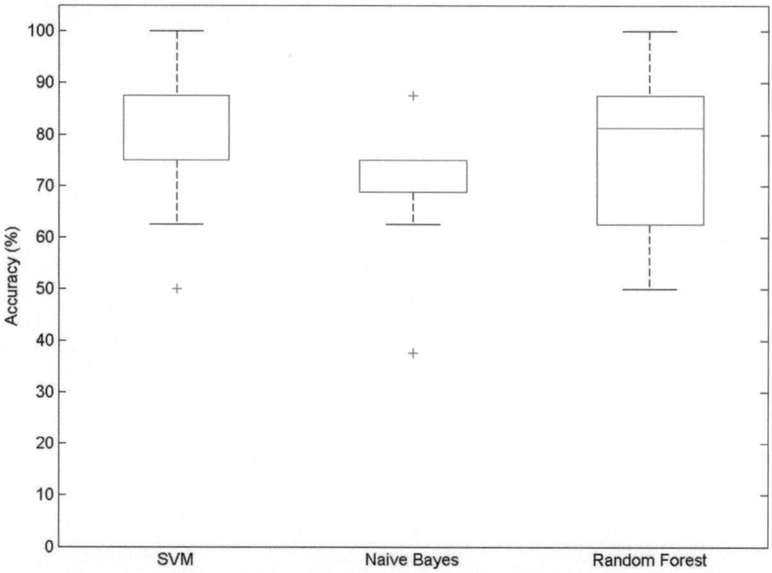

Fig. 14.3 Comparison of Accuracies using leave-one-out (2-Class)

Naïve Bayes, and Random Forest. Data was collected from 12 participants who performed an offline variation of the experiment with no real-time adaptation, using the same task design. For each participant, an equal number of tasks was performed for CL levels {0, 4, 8, 12}. Leave-one-out cross-validation across the participants was used to evaluate ML performance.

As shown in Fig. 14.3, for 2-class classification based on the collected offline GSR signals, SVM slightly outperformed the other two with the accuracy of 78.1 % compared to 71.9 % for Naïve Bayes and 76.0 % for Random Forest.

For the 4-class classification as shown in Fig. 14.4, SVM also outperformed the other two ML methods (49.0 %), compared to 40.6 % and 36.5 % for Naïve Bayes and Random Forest respectively. Therefore, SVM was chosen as the ML method during the dynamic workload adaptation.

14.4.2 Online Cognitive Load Classifications

The online cognitive load classification during the dynamic workload required a slightly different cross-validation method compared to the offline classifications. In the online cognitive load classifications, the ML model was trained using a calibrated and therefore *personalized* version of the same static data used in the offline cross-validation. After extracting the features from the data, they were calibrated using the mean and variance of the features extracted from a short training stage conducted by the particular subject. In this way, the participant did not have to

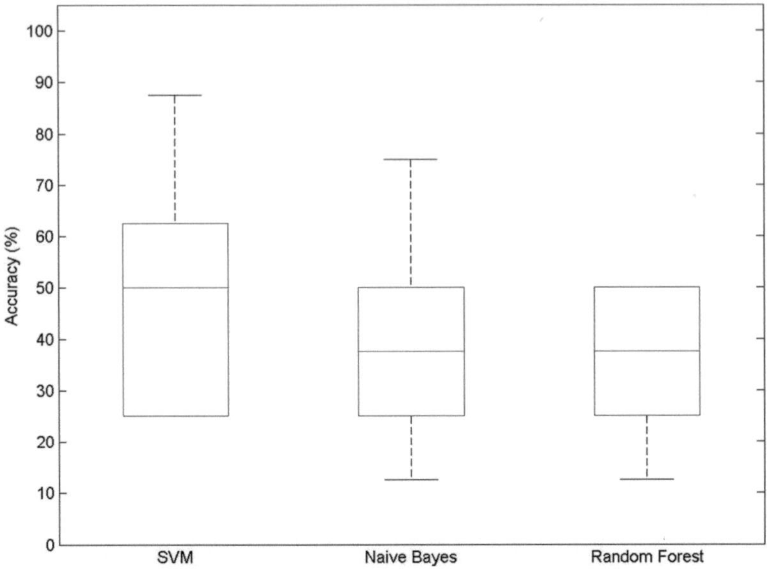

Fig. 14.4 Comparison of Accuracies using leave-one-out (4-Class)

undergo an extremely long training stage, and the classification model could still take into account the subjective differences between participants to a certain extent.

Correctness and therefore *accuracy* of the classifications also need to be more clearly defined for online classifications as the CL levels are all integer values ranging from 0 to 12 for the tasks, but classifications are only made from the set of {0, 4, 8, 12}. *Correctness* in the online 4-class classification is defined as follows: for any CL level that falls between two of {0, 4, 8, 12}, a classification is considered *correct* if it matches one of these two *neighboring* values. For example, if the CL level is 3, a classification of either 0 or 4 is considered *correct*. For a 2-class problem, it is difficult to use a similar logic, thus a *correctly* classified value was defined to be within 6 levels of the true CL level. The *correctness* of classification for the online analysis was purposely defined more loosely due to issues that arise with stricter definitions given the nature of the task design. As a consequence of the inherent randomness presented in the tasks, the CL level can only be used as a good indicator of the difficulty rather than an exact metric. This is especially true of the online classification problem involving 13 different CL levels as the margins between each level are not as distinguished. For the offline case which only involved levels {0, 4, 8, 12}, this issue is minimized as the margins between levels are less disputable.

Table 14.3 shows the comparison of classification accuracy between offline classifications and online classifications. The results show that all ML methods have similar or even better performance in online CL classifications as do offline CL classifications. All three ML methods can effectively classify CL levels of tasks in real-time.

Table 14.3 Accuracy of offline classifications vs. online classifications

	2-class (%)		4-class (%)	
Classifiers	Offline	Online	Offline	Online
SVM	78.1	81.7	49.0	44.0
Naïve bayes	71.9	89.2	40.6	63.6
Random forest	76.0	88.4	36.5	56.6

Despite having the best performance during the offline classification, SVM showed the worst mean accuracies during the online classification. The main difference between the offline and the online classifications is that the calibration of data was conducted during the online classification process in order to *personalize* the data for the given participant. These interventions may have had varying effects on different ML algorithms, and therefore produce the differences seen in the accuracy between the online and offline classifications.

14.5 Dynamic Workload Adjustment

14.5.1 Adaptation Models

The objective of the adaptation model is to keep the CL level within the range 4–8 (middle range of CL levels) during the dynamic workload adaptation. Two adaptation models are designed in this study.

- Adaptation Model 1 (AM1):
  - If classified CL level CL_t at time t is to be 8 or 12, decrease the CL level by 1;
  - If CL_t is classified to be 0 or 4, increase CL level by 1.
- Adaptation Model 2 (AM2):
  - If CL_t is classified to be 12, or CL_t and CL_{t-1} are both classified to be 8 or above, decrease CL level by 1;
  - If classified level is 0, or CL_t and CL_{t-1} are both classified to be 4 or below, increase CL level by 1.

Adaptation model 1 can be regarded as the more *dynamic* variation model, while the adaptation model 2 is clearly more *stable* as it has to meet slightly more strict criteria to change levels. The design of both adaptation models is also kept consistent with their objective by ensuring that their behaviors are symmetrical about the mean value of the desired range, ie the level 6.

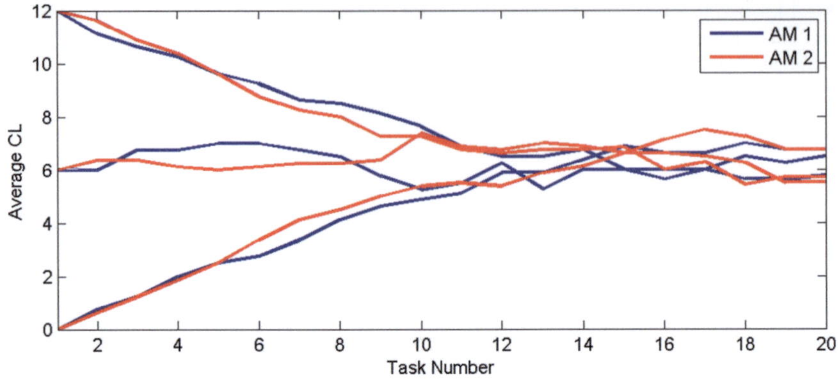

Fig. 14.5 CL Changes throughout testing stages

Table 14.4 Average performance of adaptation models

Initial CL	AM1 (%)	AM2 (%)
0	53.1	56.9
6	87.5	86.3
12	41.9	51.5

14.5.2 Performance Evaluation of Adaptation Models

Performance of the adaptation model is defined as the percentage of tasks which have $4 \leq CL \leq 8$ out of all tasks performed during a testing stage. It is regarded as the *desirable* range in the dynamic workload adaptation. Figure 14.5 shows the changes of average CL levels during the dynamic adaptation process. Each point (n, CL) on a curve represents the average CL level for the corresponding task number (n) of a particular testing stage scenario, differentiated by their initial CL and the adaptation model used. In Fig. 14.5, it is clear to see that all adaptation models were able to successfully drive and maintain the CL level within the desired range, and only minor differences being observed between the two adaptation models. For a quantitative comparison, the average performances for each scenario are summarized in Table 14.4.

Significant differences are only observed for $CL_{initial} = 12$, where adaptation model 2 achieved around 10 % higher performance than adaptation model 1. Reasons for poorer and more disparate performance levels for this scenario could be attributed to the subjectivity inherent in the experiment that could not be completely removed from the task design. Subjectivity is a more significant issue for higher difficulty level tasks, as the GSR responses are likely to show greater variations between participants due to differences in arithmetic ability. With this increased diversity, accurate classification becomes more challenging, and translates to poorer performance.

14.6 Summary

This chapter investigated the use of GSR features in dynamic workload adjustment. A dynamic workload adaptation feedback loop was presented for the dynamic workload adjustment. Both time domain and frequency domain features were extracted and used for CL level classifications. The experimental results showed that SVM, Naïve Bayes, and Random Forest were all able to provide reasonable accuracies of CL level classifications. Furthermore, the classification results could be used as inputs to CL adaptation models resulting in a dynamic workload adjustment environment, where the CL level is driven and maintained around an optimal level.

References

1. J. Zhou, J.Y. Jung, F. Chen, Dynamic workload adjustments in human-machine systems based on GSR features, in *Human-Computer Interaction – INTERACT 2015*, ed. by J. Abascal, S. Barbosa, M. Fetter, T. Gross, P. Palanque, M. Winckler (Springer International Publishing, Cham, 2015), pp. 550–558
2. P. Chandler, J. Sweller, Cognitive load theory and the format of instruction. Cogn. Instr. **8**(4), 293–332 (1991)
3. F. Paas, J.E. Tuovinen, H. Tabbers, P.W.M. Van Gerven, Cognitive load measurement as a means to advance cognitive load theory. Educ. Psychol. **38**(1), 63–71 (2003)
4. D. Afergan, E.M. Peck, E.T. Solovey, A. Jenkins, S.W. Hincks, E.T. Brown, R. Chang, R. J.K. Jacob, Dynamic difficulty using brain metrics of workload, in *Proceedings of the SIGCHI Conference on Human Factors in Computing Systems,* New York, NY, USA, 2014, pp. 3797–3806
5. A. Koenig, D. Novak, X. Omlin, M. Pulfer, E. Perreault, L. Zimmerli, M. Mihelj, R. Riener, Real-time closed-loop control of cognitive load in neurological patients during robot-assisted gait training. IEEE Trans. Neural. Syst. Rehabil. Eng. **19**(4), 453–464 (2011)
6. N. Nourbakhsh, Y. Wang, F. Chen, GSR and blink features for cognitive load classification, in *Human-Computer Interaction – INTERACT 2013*, ed. by P. Kotzé, G. Marsden, G. Lindgaard, J. Wesson, M. Winckler (Springer, Berlin/Heidelberg, 2013), pp. 159–166
7. J. Son, M. Park, Estimating cognitive load complexity using performance and physiological data in a driving simulator, in *Proceedings of AutomotiveUI'11,* Salzburg, Austria, 2011
8. W. Wang, Z. Li, Y. Wang, F. Chen, Indexing cognitive workload based on pupillary response under luminance and emotional changes, in *Proceedings of the 2013 International Conference on Intelligent User Interfaces*, 2013, pp. 247–256

Chapter 15
Real-Time Cognitive Load Measurement: Data Streaming Approach

There are many time critical problems that require detecting and managing cognitive load in real-time. In this chapter, we present a reformulation of the problem of cognitive load change detection as the problem of concept shift/drift detection in data streams. A good summary of the concept drift adaptation is given here [1]: monitoring user behavior in real-time provides us with several data streams. These data streams can individually (or collectively) be processed to detect sudden *shifts* or gradual *drifts* in behavior. This chapter explores the issues resulting from this problem reformulation and the empirical results obtained from its implementation as proposed in [2].

As discussed in previous chapters, cognitive load is the load on working memory experienced by a user when undertaking a learning or mission critical task. Managing cognitive load is desirable for enhancing productivity and avoiding fatal scenarios in time critical situations. However, measuring cognitive load remains the holy grail of cognitive load theory [3]. A number of approaches exist for measuring cognitive load but with associated limitations [4]. Subjective measures are based on a user's feedback via detailed questionnaires, whereas performance oriented measures are derived from the accuracy and precision of the task in question. Both these (post-event) approaches are popular as a ground truth in experimental conditions; however, they are not useful when cognitive load changes are to be detected in real time (or in the wild). Physiological and behavioral approaches appear more suited for real time scenarios. Both assume that higher mental effort will cause observable involuntary/voluntary changes in human physiological/behavioral measures. Some physiological measures that have delivered significant results include speech features [5] and galvanic skin response [4]. Multimodal behavioral measures (including digital pen and eye gaze) are also being actively researched [4]. In this chapter we focus on a non-intrusive interactive behavioral measure that is economical, convenient to monitor and still holds good potential for real-time cognitive load indices. We then show that learning from mouse interactivity and using sliding window techniques (initially proposed in [2]) further strengthens the multimodal framework [4] for real time cognitive load detection.

F. Chen et al., *Robust Multimodal Cognitive Load Measurement*, Human–Computer Interaction Series, DOI 10.1007/978-3-319-31700-7_15

15.1 Sliding Window Implementation

The problem of cognitive load detection from sensory data streams can be reformulated as the problem of detecting concept drift from data streams. The Sliding window technique solves, usefully, the problem of detecting changes in the user's cognitive load in live scenarios. Only when significant variations are detected in real time, can there be any intelligent user interface adaptation to accommodate any anomalous load situation. Raw data streams are (usually) to be captured and retained for only a short duration (as real time implementations impose a limit on storing incoming data). Older chunks of data must be discarded as new data arrives. Here mouse curve feature vectors are streamed into sliding windows (see Fig. 15.1). This implementation maintains separate (current and reference) windows for each feature. The decision engine is responsible for pooling various feature change detections (in customizable time intervals) and then using threshold criteria to predict an overall behavior change (i.e. flag or ignore). In case a load change is flagged, the relevant feature reference model windows are

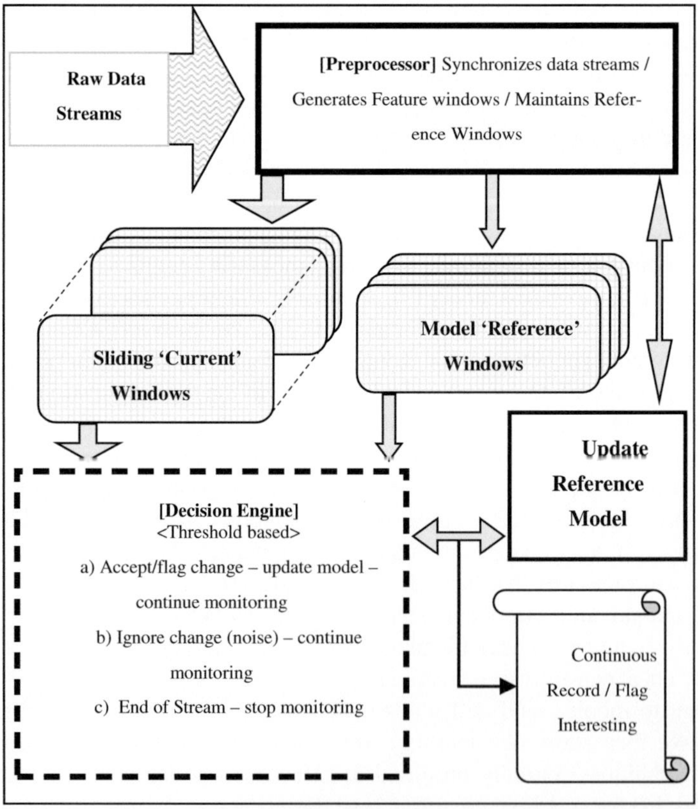

Fig. 15.1 Block diagram for sliding windows implementation

immediately updated using data from both current windows and freshly streamed points. This dynamic reference model update enables the implementation to remain relevant while continuously monitoring for changes. Every successful change detected inspires a new norm for updated model. In this particular implementation we monitored eight features using this technique. The threshold was a pooled effect of the changes detected in each feature. The 'flag condition' is interpreted to be 'set' by the arrival of five new data points, when at least two or at most six feature drifts are detected. Higher drifts of seven and eight features are treated as outliers (and in hindsight turned out to be false positives).

15.2 Streaming Mouse Activity Features

Raw mouse events (eg Move, Click, Drag etc.) are interrupt driven and can be easily monitored. These events along with system time stamps and screen coordinates are recorded. A 'pause' in mouse activity refers to the interval between two consecutive mouse events. It has been argued that both contemplation and hesitation style pause interval categories [6] hold significant promise for detecting a high cognitive load on working memory. Contemplation-style intervals correspond to breaks in user input activity that range from about 1 to 5 s and are easily observable. This type of interval exhibits change patterns similar to those previously observed for speech pause features [7]. On the other hand, hesitation-style intervals typically range from more than one-tenth of a second to one second apart. These correspond to subtle variations in user input behaviour, that may be interpreted as 'hesitant' or 'cautious' due to the high cognitive load on working memory.

This idea is further extended by extracting 'trajectories' associated with hesitation-style intervals. This moves us from the temporal to the spatial dimension. User mouse activity is currently segmented into trajectory curves, similar to those proposed by Schulz [5], but with more detailed features. The eight mouse curve features observed (and streamed) include length and number of sample points per curve; curvature (four components) and inflection (two components). These 'mouse curves' represent normalized chunks of user activity that may be streamed to online machine learning techniques for detecting changing patterns or anomalies.

A trajectory is the path followed by an object in space as a function of time. In the case of a mouse trajectory, this path is interpolated from the time stamped coordinates made available by consecutive mouse events (see Fig. 9.3). Time sampled chunks of data are used as the basis points for generating splines (representing the actual user mouse curve in 2D). Several issues are of concern here with regards to 'mouse curves'; as to how closely the interpolated splines actually represent the user's behavior. Some of the extracted features (like actual location and velocity) are direct representation of user behavior, whereas others (like curvature and inflection points) are calculated approximations. From an operations point of view, the preprocessor (see Fig. 15.1) continues to collect consecutive ('mouse-moved' type) raw data points (including screen X and Y coordinates) till a time-stamp difference of greater than 0.1 s and less than 1 s signals the completion of one mouse 'curve'. This bunch of raw observations (typically ranging from 10 to 200 points) is grouped to create one mouse 'curve'. From this 'curve', eight features (viz. number of observation points, Euclidian length of curve, positive and negative curvatures of vertical and horizontal activity components, inflexion points for both vertical and horizontal movement) are extracted into a feature vector. Several contiguous feature vectors are passed on to the decision engine as feature windows (see Fig. 15.1).

The analysis here follows on from a larger multitier experiment that was designed to study the effects of cognitive load on organizational trust. Cognitive load was varied using a standard dual-task design [8]. The low load condition comprised of a sole primary task, whereas the high load condition included an additional secondary task that periodically popped into the user's view and conditionally required a classification action. The experiment was conducted through a simulated computer-based platform to screen applicants for a human resource department. Every subject interacted with this tool; studied the information provided and completed the AIB (Ability, Integrity & Benevolence) trust indicator related task [9] while user interaction data (including mouse activity) was recorded. All subjects completed both (low and high cognitive) load conditions in a repeated measures design. The scope of the current analysis was limited to processing user mouse activity streams captured from 27 subjects. Each subject data stream was approximately balanced (in terms of data points) for both load level mouse activity. It should also be noted that these data streams lasted comparatively long periods of time (typically 20–30 min per subject per load level session). The two (high and low) load session data streams of each subject were interleaved carefully to create a single large stream (per subject) that contained 19 load drift points (for that subject). Actual load labels were then removed and a sliding window algorithm allowed process each subject activity stream for load change detection.

Figure 15.2 shows the Positive Predicted Values (PPV) of load variations detected for the 27 subject streams. Every stream was based on a ground-truth of 19 true drifts and a mean value of 8.2 true positive detections was achieved (which is reasonably good considering the fact that the reference model was updated dynamically after each detection, whereas earlier approaches had limited or no capability of learning from new data).

Table 15.1 shows spread of actual detections and true detections. One subject stream achieved 100 % accuracy. All eight changes detected in this case were true positives.

15.3 Lessons Learnt

Several aspects of the popular techniques were customized to address our problem. Standard ADWIN [8] is a popular choice, but only monitors the mean values and once a change is detected, adopts a reference window size to maintain only the new

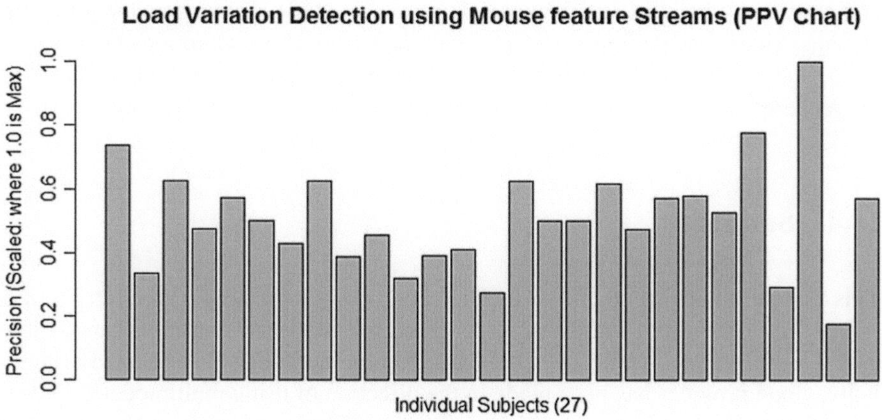

Fig. 15.2 Load variation detection (precision chart)

Table 15.1 Actual and true detections

	Min	Mean	Max
Actual detections	8	17.15	34
True detections	3	8.2	17

Detection by our technique was considered a ‘true positive’ if the flagged change by algorithm and actual change were within 5–85 % of window size. This definition results from the operational size of the current and reference windows implemented

values statistically consistent with the new null hypothesis. Since a more complex (user) behavior pattern was needed to maintain and detect cognitive load, we opted for more informative, intensive structures as depicted in Fig. 14.1. These structures pose a challenge to the limited memory and time resources available in stream scenarios, however, they have the capability to be further extended to include data streams arriving from other modalities. Asynchronous data streams are another challenge in this regard that need to be addressed in future research. A threshold is needed to maintain both the signal/noise distinction and also to accommodate the within-user behavior variability. Currently we base this threshold on theoretical data analysis, however, this threshold could typically be defined in more detail as a (complex) function based on contextual variables like user type, activity type and session duration. Typical user behavior tends to fluctuate for extended durations of activity and may also differ according to the nature of the task at hand. Any pattern/behavior beyond the accepted normal threshold is flagged as interesting while other variations are ignored. The flagged behavior is then to be further processed for warranting any suitable action.

Finally, the Kolmogorov-Smirnov Detection test that we use is quite popular in drift detection studies [9]. It was found to be suitable to the case presented in this chapter since we are interested in a nonparametric test that signals any change in distribution and not just a single aspect of the random variable. Also to be explored and considered here are the temporal dependencies [10] of points arriving in data streams. Since they no longer comply with the traditional assumptions of independence and identical distributions, a number of standard approaches and techniques are rendered ineffective or must be greatly reworked to give valid results.

15.4 Summary

This chapter stated that cognitive load detection is critical to user's productivity and safety. The efficacy of intelligent user interfaces would be greatly enhanced if a user's cognitive load could be sensed in real time and adjustments made accordingly. Good progress has been made in the direction of using multimodal behavior and interactivity as an indicator of cognitive load. These approaches were extended by enhancing the multimodal behavioral model to include mouse interactivity streams and a modified sliding window implementation. As we continue with our ongoing experimentation into modified sliding window approaches to measuring cognitive load, a clearer picture is forming as to the more general applications and limitations of real-time measurement based on these measures.

References

1. J. Gama, I. Žliobaite, A. Bifet, M. Pechenizkiy, A. Bouchachia, A survey on concept drift adaptation. ACM Comput. Surv. **46**(4), 44:1–44:37 (2014)
2. S. Arshad, Y. Wang, F. Chen, Interactive mouse stream as real-time indicator of user's cognitive load, in *Proceedings of the 33rd Annual ACM Conference Extended Abstracts on Human Factors in Computing Systems*, 2015, pp. 1025–1030
3. P. Ayres, F. Paas, Cognitive load theory: New directions and challenges. Appl. Cogn. Psychol. **26**(6), 827–832 (2012)
4. F. Chen, N. Ruiz, E. Choi, J. Epps, M.A. Khawaja, R. Taib, B. Yin, Y. Wang, Multimodal behavior and interaction as indicators of cognitive load. ACM Trans. Interact. Intell. Syst. **2**(4), 22:1–22:36 (2012)
5. D.A. Schulz, Mouse curve biometrics, in *Biometric Consortium Conference, 2006 Biometrics Symposium: Special Session on Research at the*, 2006, pp. 1–6
6. S. Arshad, Y. Wang, F. Chen, Analysing mouse activity for cognitive load detection, in *Proceedings of the 25th Australian Computer-Human Interaction Conference: Augmentation, Application, Innovation, Collaboration*, 2013, pp. 115–118
7. B. Yin, F. Chen, Towards automatic cognitive load measurement from speech analysis, in *Human-Computer Interaction, Interaction Design and Usability*, ed. by J. Jacko (Springer, Berlin, 2007), pp. 1011–1020
8. R. Brunken, J.L. Plass, D. Leutner, Direct measurement of cognitive load in multimedia learning. Educ. Psychol. **38**(1), 53–61 (2003)
9. R. Mayer, J.H. Davis, F.D. Schoorman, An integrative model of organizational trust. Acad. Manag. Rev. **20**, 709–734 (1995)
10. I. Žliobaitė, A. Bifet, J. Read, B. Pfahringer, G. Holmes, Evaluation methods and decision theory for classification of streaming data with temporal dependence. Mach. Learn. **98**(3), 455–482 (2014)

Chapter 16
Applications of Cognitive Load Measurement

Monitoring the mental state of users can profoundly improve interactions and the user experience in today's autonomous systems. Various practical aspects of a person's life including, but not limited to, user interface design, education and training, transportation (road, rail, sea or air), emergency management, would dramatically benefit from objective, robust, accurate, real-time, unobtrusive detection of cognitive load. Both behavioural and physiological methods can provide such measurement. The highlights of this chapter include:

- Cognitive load measurement in user interface design;
- Application of cognitive load measurement in emergency management;
- Cognitive load measurement in driving and plane piloting;
- Monitoring learner's cognitive load in education and training to improve the learning efficiency;
- Future applications of cognitive load related technologies.

16.1 User Interface Design

User interface design can benefit significantly from cognitive load measurement, since the representation of information (graphical versus text-based, size, colour, etc.) can have a dramatic effect on the user-computer communication [1]. Sweller [2] demonstrated that ineffective user interface designs may interfere with information acquisition by increasing the associated cognitive load. Sweller et al. [3] suggested considering the following cognitive load-related effects in interface design:

- **Split-attention effect**. This effect is expected to be reduced by physically integrating different sources of information in the design in order to lower users' needs of mental integration.

F. Chen et al., *Robust Multimodal Cognitive Load Measurement*, Human–Computer Interaction Series, DOI 10.1007/978-3-319-31700-7_16

- **Modality effect**. Different modalities, if properly combined, can help to deliver information to the user in a more effective means. It is suggested to incorporate visual and auditory components together to employ different components of working memory as addressed in Baddeley's model [4], thereby decreasing the cognitive load on any single component.

Back and Oppenheim [5] found that users would prefer an interface design requiring a relatively low cognitive load that at the same time, can reach high user satisfaction. Shi et al. [6] proposed a Cognition-Adaptive Multimodal Interface (CAMI), for a large metropolitan traffic incident and emergency management system. CAMI combines complementary concepts and tools from Cognitive System Engineering and from Cognitive Load Theory (CLT). CAMI was designed for the Traffic Incident Management (TIM) system for a metropolitan Traffic Control Centre (TCC) in Australia. In the TCC, the TIM system is deployed in a control room and is composed of about twenty different computer programs, applications and devices used for incident detection, verification, response and recovery. The TIM system is operated by Traffic Control Officers (TCOs) on a 24/7 basis to manage all road traffic incidents and emergencies. One of the most important requirements for the design of the new interface put forward by the centre is the ability to reduce human cognitive load during stressful situations so that human errors and mistakes can be reduced to minimum. CAMI is centred on the question of how to detect and model human cognitive states and cognitive load, and how to incorporate this knowledge into the design of dynamic and adaptive interfaces, so that they achieve the objective of joint human-machine optimization by maximizing system performance and at the same time minimizing human cognitive load.

Figure 16.1 shows the simplified TIM system used by TCOs without CAMI. Basically, TIM operations consist of incident detection, incident verification and incident response. At the TCC analysed in [6], TCOs use more than 15 different programs, software and devices to perform TIM operations on a 24/7 basis. Many of these programs and device controllers have different user interfaces. TCOs' cognitive work load can become very high when emergency incidents occur during peak hours every day. As shown in Fig. 16.2, CAMI significantly simplifies the TIM operations, and provides TCOs with a new user interface that can sense TCOs cognitive load and provide cognitive support adaptively. CAMI does not touch heavy back-end system of detection, verification, response, field device and personnel, shown on the left hand side of the dotted line in Fig. 16.1, instead CAMI adds in a decision-supporting middleware layer between these components and the new multimodal user interface, as shown in Fig. 16.2.

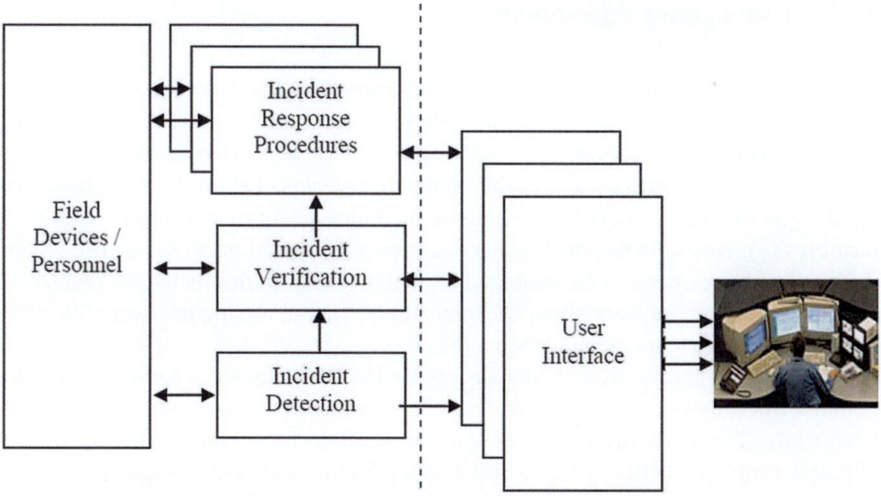

Fig. 16.1 Simplified TIM system at the TCC, without CAMI [6]

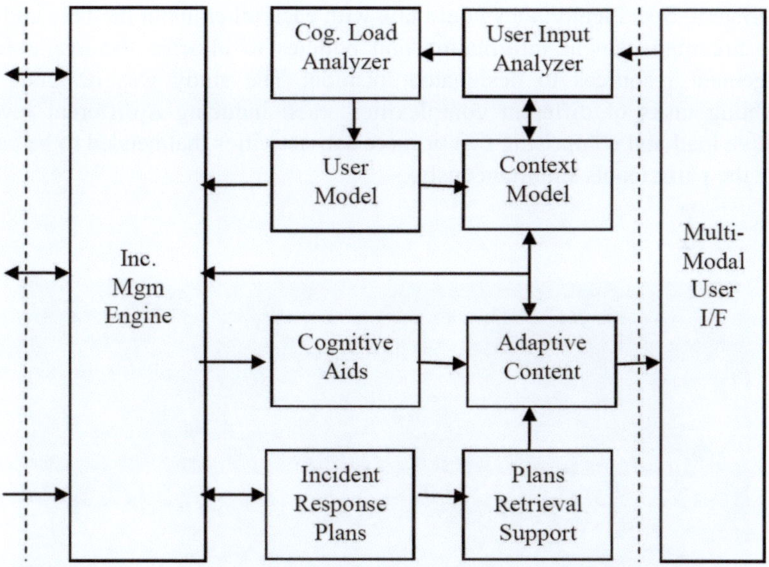

Fig. 16.2 CAMI consists of a multimodal user interface and a decision-supporting middleware that sits between the interface and the back-end TIM system (between the two *dotted lines*) [6]

16.2 Emergency Management

Emergency management is one of the extremely important domains in which a human's cognitive fatigue or mental overload can result in irreversible consequences. Ambulance service, fire-fighting, incident and crisis management are examples of such complex and highly demanding fields. Taking bushfire management as an example, in 2009, 173 people died as a result of Australia's deadliest bushfires [7]. Australia has the highest occurrences of bushfires in the world, where thousands of fires need to be managed annually. In the period between year 2002 and 2003 alone, there were about 6000 bushfires Australia-wide that were attended by the bushfire management agencies [8].

Within that perspective, Khawaja et al. [8] investigated cognitive load of humans when they were conducting bushfire emergency management via a simulated study. The study involved strategically handling fire-fighting tasks by a team of participants interacting with a multi-touch tabletop screen that displayed the fire management tasks and related information (see Fig. 16.4). The regions on the map represented different types of areas in a city (see Fig. 16.3). The participants were required to act as a group and perform complex interaction with the fire management system. Specifically, they could talk with each other about the fires and refer to the fire management information and policies to allocate the available fire management resources to designated location. The study was designed with firefighting tasks of different complexities, each inducing a different level of cognitive load and comprising two or more sub-activities that needed to be carried out by the participants simultaneously.

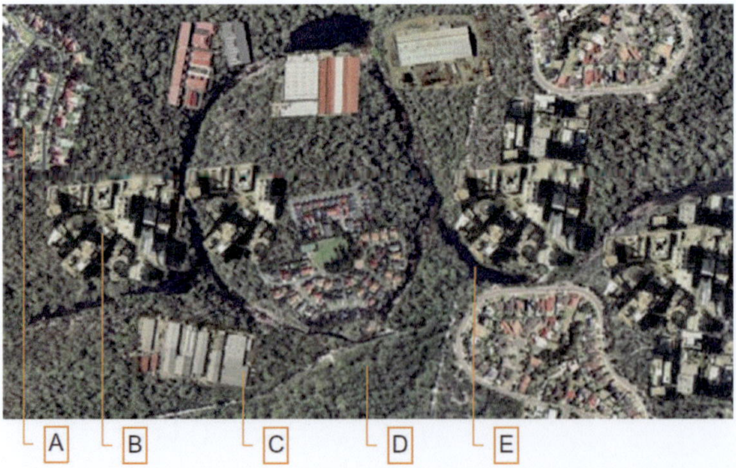

Fig. 16.3 A custom-created map of an imaginary city: (**a**) residential areas, (**b**) highrise central business districts (CBDs), (**c**) factories, (**d**) bush or parklands, and (**e**) a waterway divided the city into zones [8]

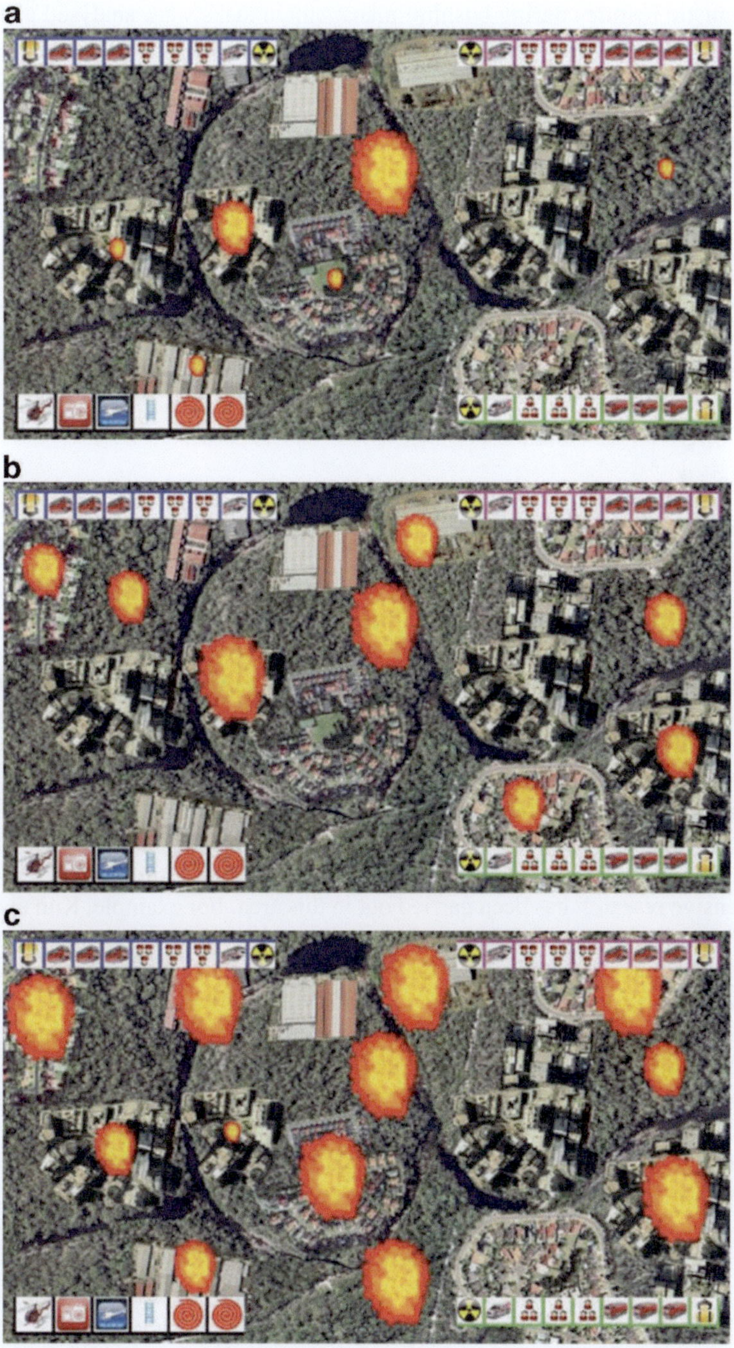

Fig. 16.4 Tasks to induce different levels of cognitive load. The severity and number of fire incidents were utilized as a factor to vary cognitive load. Screenshots of easy (**a**), medium (**b**), and hard (**c**) task levels [8]

Linguistic features during bushfire management tasks were analysed and it was found that during the emergency management participants showed significant differences in linguistic patterns between low and high cognitive load tasks [8]. The approach of using linguistic features is useful as such features can be gathered implicitly, which allows for monitoring changes in the human's linguistic patterns as they perform their work-based tasks while interacting with a system without realizing that their data are being collected and without allowing them to intentionally manipulate the data. By monitoring human's cognitive load during emergency management implicitly, human's activity can be adaptively modulated in order to improve the working efficiency.

Other emergency management domains such as traffic management and call centres can also benefit from monitoring a human's cognitive state and therefore avoiding the mistakes caused by their mental overload.

16.3 Driving and Piloting

Driver distraction via secondary in-vehicle activities is increasingly recognized as a significant source of injuries and fatalities on the roadway [9], while at the same time in-vehicle spoken dialogue systems are gaining increased interest by the automotive industry. A spoken dialogue system is a computer agent that interacts with human by understanding spoken language and provides feedback accordingly [10]. In-vehicle spoken dialogue systems enable drivers to interact with car entertainment devices, phone systems and similar while conducting the primary task of driving. Because of the dual-task paradigm for drivers using in-vehicle spoken dialogue systems, it is crucial to take the cognitive load of the driver into consideration, in order to be able to adapt the dialogue system accordingly.

Extensive research has been carried out in this area. For example, Kun et al. [10] used CLM techniques in spoken dialogue systems for interactions with in-vehicle devices to evaluate which system behaviors might result in extra driver cognitive load, which in turn could have negative safety consequences. Physiological measurement of pupil dilation was used as an indicator of cognitive load changes during dialogue while driving. It was found that the rising of pupil diameter was related to increased cognitive activity by the driver when they were attempting to find the word described by the other participant during conversion. Engstroem and colleagues [11] examined the effects of visual cues and cognitive load on driving performance and driver state during simulated and realistic motorway driving. In their study, cognitive load resulted in decreased lane keeping and increased gaze concentration toward the road centre. In another study, Crundall and Underwood [12] investigated the effects of cognitive load on the driving performance of expert and novice drivers on different types of roads. They found that experienced drivers chose visual strategies according to the complexity of the road but the strategies of novices were not flexible enough to respond properly to the variations of cognitive demand provided by each road.

Several studies utilizing simulated driving contexts have been conducted at NICTA to investigate the effect of cognitive load, its relationship with user experience, and its implications on the driving performance [13, 14]. It has been identified that high cognitive load can be detrimental to both the subjective experience and the overall driving performance, resulting in increased response time to changed road conditions, and higher risk of on-road accidents. As a consequence, the driver's cognitive load should be well monitored, and a notification mechanism should be allowed when extreme cognitive load is detected.

Besides car driving, cognitive load measurement holds promise as being a useful tool in monitoring the mental states of cockpit crews, air-craft maintenance teams, and military teams where performance is paramount, and especially for pilots, where the accuracy and speed of communication and operation is critical to safety and where errors may result in the loss of life [15]. The amount of information that needs to be processed by a pilot can be very time critical, and hence large amounts of training is expected to be conducted in simulated environments, which enables longitudinal observations of cognitive load, as well as identifying individually challenging events and providing focused training accordingly. A well-trained pilot with a good track record of cognitive load in simulated environments is a pre-requisite for safe flying in realistic piloting tasks, however cognitive load monitoring implemented in the cockpit would still be highly desirable, due to the fact that realistic pilot information could be gathered, analysed and compared with the performance of pilots in simulation training. The cognitive load of the pilots, if monitored in real time, could be used for mental state monitoring, and necessary interventions or assistance can be administered before the occurrence of critical situations.

Similar with many other application contexts, unobtrusiveness is a common requirement for cognitive load measurement both in cars and in the airplane cockpits, as any additional user requirements may result in extra cognitive load and cause usability issues, which should be minimized in any task-critical settings. As discussed in previous chapters, cognitive load can be measured via speech, eye movements, hand gestures, GSR signals etc. As technologies develop many sensors that are capable of collecting this information have been devised and embedded into wearable devices, so that these signals can be collected even without user awareness. For example, the hand gestures and BVP signals can be collected using a wristband which is smaller than the size of a watch. We can foresee that in the near future cognitive load measurement devices will be more closely geared to mission-critical environments.

16.4 Education and Training

The limited capacity of working memory based on CLT is a widely recognized determinant of human learning. Learning is different from performing a task in that new knowledge will be acquired in the process. According to CLT, learning is essentially the act of organizing information from working memory, converting it to knowledge in the form of schemas and storing the formed schemas into long term memory. Cognitive loads that exceed the capacity of working memory hamper the construction of schemas during learning. As a consequence, understanding a learner's cognitive load is critical for us to find cues to optimise learning materials, control the learning process, select the learning tasks and thus improve overall learning efficiency. The following examples will demonstrate how cognitive load measurement and monitoring methods can benefit learning.

Hessler and Henderson [16] applied CLT to nursing education. This study showed the feasibility of using cognitive load measurements to improve retention of nursing curricula information that the students were expected to learn. Young et al. [17] showed that CLT can contribute to medical education particularly because many of the medical professional activities to be learned require the simultaneous integration of multiple and varied sets of knowledge, skills and behaviours at a specific time and within a specific context. These activities feature with high "element interactivity" and therefore impose a cognitive load that may surpass the working memory capacity of the learner. Various medical education settings for CLT applications including classroom, workplace and self-directed learning can be applied and are believed to be beneficial for students to acquire nursing knowledge in a more efficient way.

Another interesting application is proposed to improve program coding skills via cognitive load examination technologies. According to Harms [18], cognitive load information is helpful for generating personalized tutorials to improve the effectiveness of learning new programming concepts found in unfamiliar source code. Specifically, they showed that there were two means to improve the effectiveness of personalized tutorials: (1) tracking the users' intrinsic cognitive load by modeling their programming expertise, and (2) reducing extraneous cognitive load by carefully selecting programming concepts that do not overwhelm a learner's working memory. Anvari et al. [19] used CLM to identify students that were talented in three-dimensional computer graphics programming via a spatial ability test. In this case a dual task approach was used for CLM. Performance-based measurement, physiological measurement and subjective surveys were used as cognitive load measures. It was found that students with high spatial ability performed better during the task of generating three-dimensional computer graphics with lower cognitive load. Those students were, as a result, identified as talented students in three-dimensional computer graphics programming. Gillmor et al. [20] used CLT in a mathematics test to improve student performance by reducing the cognitive load of math assessment items.

Cognitive load can also be manipulated by tailoring instructional design to levels of a learner's prior knowledge [21]. In this context, the instructional design is not only aimed at controlling the cognitive load, but also at stimulating learners to use their available cognitive capacity for better learning. Sweller [22] suggested that encouraging learners to exert more cognitive effort is another measure which is crucial to the construction of schemata. Learners should be motivated to devote more cognitive effort to schema construction and automation to improve their cognitive task performance by employing CLT in instructional design [21].

Coyne et al. [23] has discussed how the fields of augmented cognition and neuroergonomics can be expanded into training contexts. Several classification algorithms based on EEG data were discussed in terms of their ability to classify the human mental state in real time. These indices have been shown to enhance human performance within adaptive automation paradigms. Due to the limited capacity of the human working memory system, when the training requirement exceeds the working memory capacity, learning outcomes can be affected. Coyne et al. [23] also discussed how CLT can be combined with multiple resource theory to create an adaptive training model, which hypothesizes that a system that can monitor working memory capacity in real time and adjust training difficulty can improve learning.

16.5 Other Applications

Cognitive load measurement can be used in many other applications. For example, recent research has shown that cognitive load has an effect on gait [24], especially noticeable in human with neurodegenerative disorders such as Alzheimer's disease vascular dementia, mixed dementia [25] or Parkinson's disease [26]. More specifically, modifications in gait are associated with a decrease in the frontal cerebral blood flow [27]. For example, cerebral vascular abnormalities are associated with modifications of the gait pattern, namely an increased variability of spatio-temporal gait parameters [28]. These observations are consistent with studies claiming that gait requires cognitive processes such as attention, memory and planning [29], demanding frontal and parietal activity in the brain [24, 30]. By examining the effect of cognitive load on gait, the states of neurodegenerative disorders can be evaluated and monitored. The application of cognitive load measurement technologies can be very flexible, and one interesting example is the use of cognitive load measurement methods for assessment of newly-hired staff, aiming to identify whether a staff member is unfit or fit for a position. By examining the cognitive load of candidate staff when they are conducting tasks designed to be similar to realistic ones, their cognitive load tracking records are able to reflect their capability to fulfil the requirement of the position and their profile to learn the skills to conduct the task. A cognitive load-based selection mechanism may significantly improve the efficiency of human resource departments, and more importantly, provide an objective means to assess people's abilities with low cost.

Cognitive load measurement also plays significant roles in improving information retrieval system design [31]. In examining user interface designs of information retrieval systems, Hu et al. [32] utilized cognitive load as a measurement of the information seeking and processing effort. Utilising cognitive load measurement, they examined how searchers facilitated information gain by reducing cognitive load to increase searchers' satisfaction. Additionally, cognitive load measurement has also been applied in simultaneous interpreting tasks to improve an interpreter's performance by monitoring cognitive load in real-time [33].

16.6 Future Applications

Cognitive load knowledge and related technologies promise to help us to understand ourselves better, improve our capabilities, avoid mistakes in tasks, and improve our overall experience when interacting with the external world. The example applications discussed in this chapter have shown the utility of cognitive load examination techniques up to date, however it should be noted that many more novel applications can be devised as technologies, especially advanced hard-ware, are designed and developed.

So far many cognitive load measurement methods, although proved effective in the laboratory environment, still require cumbersome equipment which are intrusive to the users, or even restrict the movement of people when they are used. For example, to collect the EEG signal, a helmet or special cap is used – normally at least 10–15 min are required for the participant to don the helmet, and for the experiment coordinator to connect the wires and to tune the signals. Furthermore, the signal is so sensitive that subjects wearing the EEG helmet are not allowed to change their body posture abruptly, which may result in large amount of noise and overwhelm the desired EEG signals. Obviously such a device is still some way from realistic applications, however for some other cognitive load measurement methods, we can see horizon bright immediate future. Similar to EEG devices, GSR devices used to require complex set-up procedure, and allowed for minimal hand movement if they were attached to the fingers, however recent technologies have made it possible for some wristbands to monitor the GSR signal in real time, which is an encouraging step towards real-life cognitive load measurement. Another example is eye tracking glasses, which used to be heavy and illumination sensitive. New multimedia solutions, such as Google glasses and the Microsoft HoloLens are expected to feature the eye-tracking capabilities, which will make cognitive load measurement even more convenient.

With the pervasive cognitive load measurement technologies available, many aspects of our future life will be changed. We will be able to monitor our cognitive load ourselves if interested, or an automated agent will take care of our cognitive load for us. Whenever the cognitive load is beyond a threshold, appropriate interventions will be conducted automatically. For example, when students work on a mathematic problem that is too difficult for their current level, a cognitive load adaptive agent will suggest other problems for the students to start with. Personalization is another important perspective of cognitive load applications, and individual differences in cognitive load profile will be accounted for in the measurement, adaptation and recommendations.

Remote cognitive load measurement can be another important application, especially for jobs with high-risk profiles. In particular, cognitive load aware devices have attracted extensive interest from the military, and understanding the cognitive load of solders in a battlefield will definitely improve the communication and collaboration in the life-critical context. People operating complex devices or with high physical load, such as astronauts or long-journey heavy vehicle drivers, may also suffer from cognitive overload. A remote cognitive load monitoring system from the command center will allow for messages such as 'take a rest', or 'no more work today' to deliver to people who have been mentally overloaded, to assist them in managing load and avoiding potential accidents or mis-operations.

Cognitive load research, as addressed in this book, forms solid links between the internal mental world of human and the external world involving tasks, environments, context etc. In consequence, the applications listed in this chapter are just a few examples that embody the idea of cognitive load – and many more applications dependent on cognitive load technologies, as long as they are mental-status sensitive, can be added to the list. Future progress in cognitive load research methods and devices will move us closer to a deeper understanding of cognitive load mechanisms, achieve more accurate and robust measurements, and nurture more interesting applications.

References

1. N. Nourbakhsh, *Machine Learning Methods for Multimodal Cognitive Load Measurement* (The University of Sydney, Sydney, 2015)
2. J. Sweller, Cognitive load theory, learning difficulty, and instructional design. Learn. Instr. **4** (4), 295–312 (1994)
3. J. Sweller, J. Merrienboer, F. Paas, Cognitive architecture and instructional design. Educ. Psychol. Rev. **10**(3), 251–296 (1998)
4. A.D. Baddeley, Working memory. Science **255**, 556–559 (1992)
5. J. Back, C. Oppenheim, A model of cognitive load for IR: implications for user relevance feedback interaction, Inf. Res. **6**(2) (2001)
6. Y. Shi, E. Choi, R. Taib, F. Chen, Designing cognition-adaptive human–computer interface for mission-critical systems, in *Information Systems Development*, ed. by G.A. Papadopoulos, W. Wojtkowski, G. Wojtkowski, S. Wrycza, J. Zupancic (Springer, Paphos, 2010), pp. 111–119
7. Victoria Police, *Bushfires Death Toll Revised to 173* (2009). [Online]. Available: http://www.police.vic.gov.au/content.asp?Document_ID=20350
8. M.A. Khawaja, F. Chen, N. Marcus, Measuring cognitive load using linguistic features: Implications for usability evaluation and adaptive interaction design. Int. J. Hum. Comput. Interact. **30**(5), 343–368 (2014)
9. D.L. Strayer, J.M. Cooper, J. Turrill, J. Coleman, N. Medeiros-Ward, F. Biondi, *Measuring Cognitive Distraction in the Automobile* (AAA Foundation for Traffic Safety, Washington, DC, 2013)
10. A.L. Kun, Z. Medenica, O. Palinko, P.A. Heeman, Utilizing pupil diameter to estimate cognitive load changes during human dialogue: A preliminary study, in *AutomotiveUI 2011 Adjunct Proceedings*, Salzburg, Austria, 2011
11. J. Engström, E. Johansson, J. Östlund, Effects of visual and cognitive load in real and simulated motorway driving. Transport. Res. F: Traffic Psychol. Behav. **8**(2), 97–120 (2005)

12. D.E. Crundall, G. Underwood, Effects of experience and processing demands on visual information acquisition in drivers. Ergonomics **41**(4), 448–458 (1998)
13. A. Hess, J. Jung, A. Maier, R. Taib, K. Yu, B. Itzstein, *Elicitation of Mental States and User Experience Factors in a Driving Simulator* (IEEE, Gold Coast, 2013), pp. 43–48
14. J. Jung, A. maier, A. Gro, et al, Investigating the effect of cognitive load on UX: A driving study, in *Proceedings of the 3rd International Conference on Automotive User Interfaces and Interactive Vehicular Applications*, 2011, pp. 134–137
15. J.A. Cannon-Bowers, E. Salas, S. Converse, Shared mental models in expert team decision making, in *Individual and Group Decision-Making: Current Issues*, ed. by J. Castellan (Lawrence Erlbaum Associates, Hillsdale, 1993), pp. 221–246
16. K.L. Hessler, A.M. Henderson, Interactive learning research: Application of cognitive load theory to nursing education. Int. J. Nurs. Educ. Scholarsh. **10**(1), 133–141 (2013)
17. J.Q. Young, J. Van Merrienboer, S. Durning, O. Ten Cate, Cognitive load theory: Implications for medical education: AMEE Guide No. 86. Med. Teach. **36**(5), 371–384 (2014)
18. K.J. Harms, Applying cognitive load theory to generate effective programming tutorials, in *Proceedings of IEEE Symposium on Visual Languages and Human-Centric Computing (VL/HCC) 2013*, 2013, pp. 179–180
19. F. Anvari, H.M.T. Tran, M. Kavakli, Using cognitive load measurement and spatial ability test to identify talented students in three-dimensional computer graphics programming. Int. J. Inf. Educ. Technol. **3**, 94–99 (2013)
20. S. Gillmor, J. Poggio, S. Embretson, Effects of reducing the cognitive load of mathematics test items on student performance. Numeracy **8**(1) (2015)
21. S. Kuldas, L. Satyen, H. Ismail, Greater cognitive effort for better learning: Tailoring an instructional design for learners with different levels of knowledge and motivation. Psychologica Belg. **54**(4), 350–373 (2014)
22. J. Sweller, P. Ayres, S. Kalyuga, *Cognitive Load Theory* (Springer, New York, 2011)
23. J.T. Coyne, C. Baldwin, A. Cole, C. Sibley, D.M. Roberts, Applying real time physiological measures of cognitive load to improve training, in *Foundations of Augmented Cognition. Neuroergonomics and Operational Neuroscience*, ed. by D.D. Schmorrow, I.V. Estabrooke, M. Grootjen (Springer, Berlin/Heidelberg, 2009), pp. 469–478
24. E. Martin, R. Bajcsy, Leveraging wireless sensors and smart phones to study gait variability, in *Informatics Engineering and Information Science*, ed. by A.A. Manaf, S. Sahibuddin, R. Ahmad, S.M. Daud, E. El-Qawasmeh (Springer, Berlin/Heidelberg, 2011), pp. 95–111
25. G. Allali, F. Assal, R.W. Kressig, V. Dubost, F.R. Herrmann, O. Beauchet, Impact of impaired executive function on gait stability. Dement. Geriatr. Cogn. Disord. **26**(4), 364–369 (2008)
26. G. Yogev, M. Plotnik, C. Peretz, N. Giladi, J.M. Hausdorff, Gait asymmetry in patients with Parkinson's disease and elderly fallers: When does the bilateral coordination of gait require attention? Exp. Brain Res. **177**(3), 336–346 (2007)
27. T. Nakamura, K. Meguro, H. Yamazaki, H. Okuzumi, A. Tanaka, A. Horikawa, K. Yamaguchi, N. Katsuyama, M. Nakano, H. Arai, H. Sasaki, Postural and gait disturbance correlated with decreased frontal cerebral blood flow in Alzheimer disease. Alzheimer Dis. Assoc. Disord. **11**(3), 132–139 (1997)
28. C. Rosano, J. Brach, S. Studenski, W.T. Longstreth, A.B. Newman, Gait variability is associated with subclinical brain vascular abnormalities in high-functioning older adults. Neuroepidemiology **29**(3–4), 193–200 (2007)
29. D. Joshi, S. Anand, Cyclogram and cross correlation: A comparative study to quantify gait coordination in mental state. J. Biomed. Sci. Eng. **03**(03), 322–326 (2010)
30. P.L. Sheridan, J.M. Hausdorff, The role of higher-level cognitive function in gait: Executive dysfunction contributes to fall risk in Alzheimer's disease. Dement. Geriatr. Cogn. Disord. **24**(2), 125–137 (2007)
31. K. Na, Exploring the effect of cognitive load on the propensity for query reformulation behavior. PhD thesis, The Florida State University, 2012

32. P.J.-H. Hu, P.-C. Ma, P.Y.K. Chau, Evaluation of user interface designs for information retrieval systems: A computer-based experiment. Decis. Support. Syst. **27**(1–2), 125–143 (1999)
33. K.G. Seeber, Cognitive load in simultaneous interpreting: Measures and methods. Target **25** (1), 18–32 (2013)

Part VI
Conclusions

Chapter 17
Cognitive Load Measurement in Perspective

Cognitive load is a significant factor in various application areas such as HCI, adaptive automation and training, traffic control, performance prediction, driving safety, and military command and control. Consequently, the investigation of cognitive load factors and cognitive load measurement is essential in order to improve human's wellbeing and safety at work. As indicated in Part I, an individual human has limited cognitive resources. Both theories and models have been proposed to understand and measure cognitive load. Cognitive load theory models the interaction between limited working memory and the relatively unlimited long term memory during the learning process. It distinguishes between three types of cognitive load: intrinsic load, extraneous load, and germane load. The first type is associated with the nature of learning material, while the latter two are influenced by instructional design. Techniques used for cognitive load measurement can be divided into the following categories: subjective ratings, performance measures, behavioral measures and physiological measures. The physiological approaches for cognitive load measurement are based on the assumption that any changes in human cognitive functioning are reflected in the human physiology, and have attracted increasing attention. Popular physiological measures used in cognitive load studies include brain wave, eye activity, respiration, heart rate, skin conductance, and speech, etc. Response-based behavioral features for cognitive load measurement are those that can be extracted from any user activity that is predominantly related to deliberate/voluntary task completion. This book focuses on the use of the following modalities for cognitive load measurement: eye, skin conductance, digital pen, speech, linguistic, mouse activity as well as fusions thereof.

Multimodal cognitive load measures have elements derived from both human factors studies and psychology of learning. These elements, in the context of multimodal interfaces, provide the robust MCLM framework that is applicable to real-time problem scenarios.

Eye-based measures and GSR-based measures are two widely used physiological indexes for cognitive load measurement as discussed in Part II. Pupillary response features such as mean-difference pupil size were suggested for cognitive

F. Chen et al., *Robust Multimodal Cognitive Load Measurement*, Human–Computer Interaction Series, DOI 10.1007/978-3-319-31700-7_17

load measurement under the influence of luminance condition. Boosting algorithms were used for pupillary feature selection and workload classification. GSR features for cognitive load measurement from two different experiments were also demonstrated, which are temporal and spectral features of GSR. It was shown that accumulative GSR, the power spectrum of GSR, blink numbers and blink rate are significantly distinctive and have reasonable accuracies in both two- and four-class classification of cognitive load using support vector machines and Naïve Bayes classifiers.

Besides physiological measures, cognitive load can be examined via behavioral features, such as speech, linguistics and handwriting as discussed in Part III. Specifically, cognitive load affects the source from which speech signals originate, the pathway that speech goes through, and the way it is uttered. Furthermore, the usage of words, the rhythm of speaking as determined by pauses, and the emotion of speech, detected via linguistic features, are all effective features for cognitive load investigation. Overall, high cognitive load affects speech in two respects: it affects the way speech is generated, and the performance of the speech, with distinguishable linguistic implications.

On the other hand, cognitive load has a significant effect on many aspects of handwriting: the writing speed, pressure of the pen tip and the way pen is grasped are all affected by cognitive load. This has shown that writing as a learned skill, even when carried out with high levels of expertise, still requires mental resources, although possibly just a small amount, to control the required gestures. A high cognitive load can drain the mental resources and hence affect the behavior and performance of writing, and this effect can be identified from differing handwriting recognition rates associated with cognitive load variations.

Similarly, user mouse interactivity is also an important component of user input behavior. Under high cognitive load both temporal and spatial features of mouse interactivity show significant change in patterns. Mouse trajectories (or path curves) and usage of button controls both seem to be fairly good predictors of the changing cognitive load conditions of the user.

Multimodal interaction continues to be the preferred mode in the world of interface design. Given the success in finding features from single modalities that allow us to differentiate cognitive load levels, the next step is to apply a multimodal index of load that combines output from different sources. An abstract model for multimodal assessment was presented in Part IV to demonstrate feasibility of multimodal cognitive load measurement. We also showed how in practice, various confounding factors unrelated to workload, including changes of luminance condition and emotional arousal might degrade workload measures such as the commonly used mean pupil diameter. Part IV investigated pupillary response and GSR as a cognitive load measure under the influence of such confounding factors. The mean-difference feature and its extension (Haar-like features) were used to characterize physiological responses of cognitive load under luminance and emotional changes. Besides luminance condition and emotional arousal, Part IV also investigated the effect of stress on cognitive load measurement using GSR as a physiological index of CL. It was found that without the impact of stress, it appears that an increase in CL (induced by increasing the difficulty of tasks given to test subjects) results in an increase in mean GSR values. This relationship is, however, obfuscated when test subjects experience fluctuating levels of stress. More interestingly, Part IV also discussed the relations between trust and cognitive load. Multimodal features were analyzed to find how trust perception affects cognitive load and vice versa.

In order to make cognitive load measurement accessible, Part V discussed two aspects of cognitive load measurement: dynamic cognitive load adjustment and real-time cognitive load measurement via data streaming. A dynamic workload adjustment feedback loop was presented intended to control workload during human-machine interactions. In this model, physiological signals such as GSR are utilized to obtain passive human sensing data. By analyzing the obtained sensing data in real-time, task difficulty levels are adaptively adjusted to better fit the user capacities during work tasks.

With real-time data streaming the efficacy of intelligent user interfaces becomes greatly enhanced as a user's cognitive load can be instantly sensed and adjustments made accordingly. Real-time cognitive load detection using data streams was modelled to detect sudden *shifts* or gradual *drifts* in behavior. This model made use of mouse interactivity data streams and a modified adaptive sliding windows technique and demonstrated reasonable success. Some typical applications of cognitive load measurement were also investigated in Part V to demonstrate the feasibility of cognitive load measurement in practical applications.

We imagine that future work on MCLM will include furthering the cause of Bayesian models [1–3] of cognition (as well as the more recently introduced Quantum models [4] of cognition) by providing them with a framework that employs empirical/sensory links to external tangible world. Such theoretical models of cognition hold much promise for modelling the possible states of the mind and the transitions between them. Data-driven MCLM can provide empirical associations of such mental states and corresponding transformations to various user interaction modality patterns.

Although aiming at a comprehensive coverage of cognitive load measurement methods, this book is just one solid step towards robust, real-time and reliable cognitive load evaluation. Many current techniques, such as EEG, GSR, digital pen, eye tracker etc. have been applied in the examinations of this book, however there is still much space to explore and much to improve on for cognitive load measurement, likely via employing new technologies and devices, such as virtual reality methods for experiments, big data analytics for group experiences of cognitive load, longitudinal tracking to construct cognitive profile for a single subject, and experience-based cognitive load studies. The summary of the techniques presented in this book, along with their evaluation, we hope has shed light on the enormous possibilities presented by cognitive load measurement, and can now be used as a platform for the continual quest for more user-centred, adaptive technological systems.

References

1. N. Chater, J.B. Tenenbaum, A. Yuille, Probabilistic models of cognition: Conceptual foundations. Trends Cogn. Sci. **10**(7), 287–291 (2006)
2. T. Griffiths, C. Kemp, J. Tenenbaum, Bayesian models of cognition, in *The Cambridge Handbook of Computational Psychology*, ed. by R. Sun (Cambridge University Press, Cambridge, 2008)
3. J.L. Austerweil, S.J. Gershman, T.L. Griffiths, Structure and flexibility in Bayesian models of cognition, in *The Oxford Handbook of Computational and Mathematical Psychology*, ed. by J.T. Townsend, J.R. Busemeyer (Oxford University Press, Oxford, 2015)
4. P.D. Bruza, Z. Wang, J.R. Busemeyer, Quantum cognition: A new theoretical approach to psychology. Trends Cogn. Sci. **19**(7), 383–393 (2015)

MIX
Papier aus verantwortungsvollen Quellen
Paper from responsible sources
FSC® C105338

If you have any concerns about our products,
you can contact us on
ProductSafety@springernature.com

In case Publisher is established outside the EU,
the EU authorized representative is:
Springer Nature Customer Service Center GmbH
Europaplatz 3, 69115 Heidelberg, Germany

Printed by Libri Plureos GmbH
in Hamburg, Germany